Mowing and Spraying Equipment

FUNDAMENTALS OF SERVICE

PUBLISHER

Fundamentals of Service (FOS) is a series of manuals created by Deere & Company. Each book in the series was conceived, researched, outlined, edited, and published by Deere & Company. Authors were selected to provide a basic technical manuscript which could be edited and rewritten by staff editors.

PUBLISHER: DEERE & COMPANY SERVICE TRAINING, Dept. F, John Deere Road, Moline, Illinois 61265; Director of Service: Robert A. Sohl.

SERVICE TRAINING EDITORIAL STAFF
Managing Editor: Louis R. Hathaway
Editor: Laurence T. Hammond
Editor: Alton E. O'Banion
Promotions: Annette M. LaCour

TO THE READER

PURPOSE OF THIS MANUAL

The main purpose of this manual is to explain mowing and spraying equipment. Starting with how it works, we build up to why it fails, and what to do about it. This manual is a good reference for the experienced serviceman who wants to refresh his memory and the new man just learning. It is written with many illustrations so that it can be understood easily.

WHAT IS "FUNDAMENTALS OF SERVICE"?

This manual is part of a series of texts and visuals called "Fundamentals of Service," (FOS). These materials are basic information for use by teachers as well as shop servicemen and the layman. All kinds of equipment are covered — automotive and off-the-road. Emphasis is on theory of operation, diagnosis, and repair.

OTHER MANUALS IN THIS SERIES

Other manuals in the FOS series are:

- **Hydraulics**
- **Engines**
- **Electrical Systems**
- **Power Trains**
- **Shop Tools**
- **Welding**
- **Air Conditioning**
- **Tires and Tracks**
- **Belts and Chains**
- **Bearings and Seals**
- **Fiberglass/Plastics**
- **Fuels, Lubricants, and Coolants**
- **Identification of Parts Failures**

Each manual is backed up by a set of 35 mm color slides for classroom use.

FOR MORE INFORMATION

Write for a free *FOS Catalog of Manuals and Visuals*. Send your request to:

John Deere Service Training
Dept. F
John Deere Road
Moline, Illinois 61265

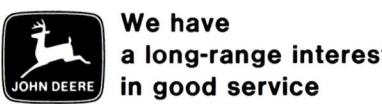

We have a long-range interest in good service

ACKNOWLEDGEMENTS

John Deere gratefully acknowledges help from the following groups: American Oil Co.; ARCO Chemical Co.; Carver Pump Co.; Delavan Mfg. Co.; Cebageigy Chemical Corp.; Lindsay Mfg. Co.; Monsanto Co.; Stauffer Chemical Co.

Copyright © 1971, 1974, 1980, 1984 / DEERE & COMPANY, Moline, Ill. / Fourth Edition / All rights reserved.

FOS — 56   Litho in U.S.A.

CONTENTS

Part 1 — MOWING CUTTER BARS

Introduction	1
Kinds Of Cutter Bars	2
Cutter Bar Parts	3
Knife	3
Maintenance and Adjustment	5
Adjusting The Cutter Bar	6
Adjusting The Knife	7
Adjusting The Guards	9
Adjusting Cutter Bar Lead	11
Adjusting The Operating Height	12
Adjusting The Cutter Bar Tilt	13
Repair and Replacement	14
Removing The Knife	14
Repairing The Knife	15
Servicing The Guards	18
Servicing The Shoes	20
Rotary Knife Cutter Bars	21
Summary	22
Test Yourself	22

Part 2 — SPRAYING NOZZLES

Introduction	23
Flow Rate	24
Atomization	24
Types Of Nozzles	25
Parts Of A Nozzle	26
Nozzle Body and Cap	26
Nozzle Strainers	27
Nozzle Tips and Spray Patterns	27
Other Types Of Spray Equipment	30
Selection Of Nozzles	30
Calibrating The Application Rate	30
Pre-Calibrating	31
How To Calibrate	32
Nozzle Adjustment and Care	33
Problems With Nozzles	33
Cleaning Nozzles	34
Test Yourself	34

FOS — 56 Litho in U.S.A.

MOWING CUTTER BARS / PART 1

Fig. 1 — Mower Cutter Bar In Operation

INTRODUCTION

Mowing cutter bars cut grass, hay, and other crops by shearing them off with reciprocating knives. Cutter bars are used on mowers, grain combines, forage harvesters, and windrowers.

Knife sharpness and proper adjustment are the most important factors to keep in mind. Cutter bar operation, types, care, adjustment, repair, and maintenance will be explained here.

A cutter bar cuts much like hedge trimming shears (Fig. 2). The knife on a cutter bar is pushed back and forth over the guard surfaces, cutting with each stroke (Fig. 3).

2 Mowing Cutter Bars

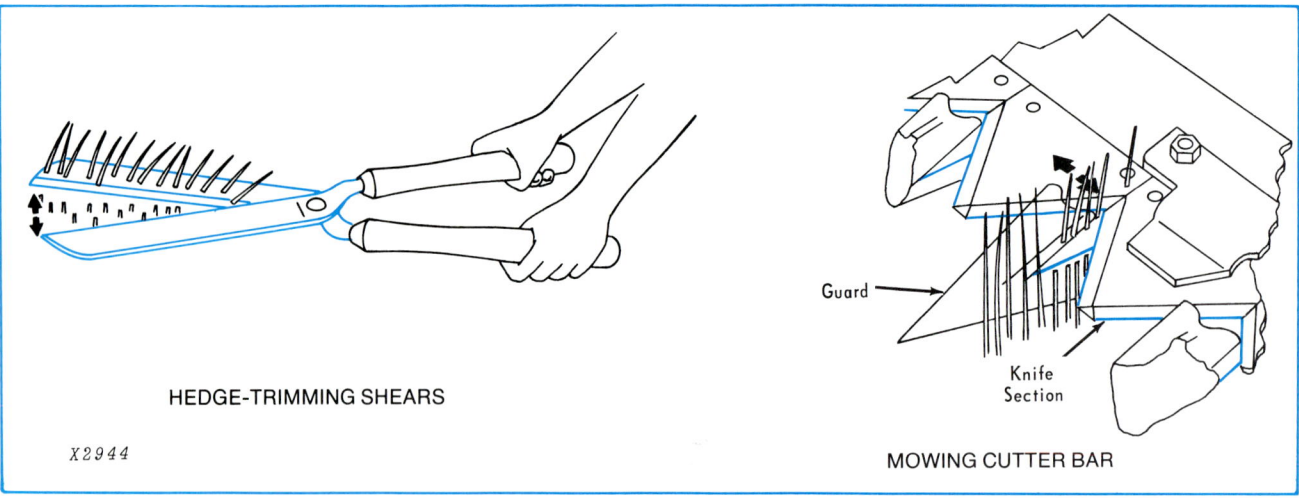

Fig. 2 — Mowing Cutter Bars Operate On The Same Principle As Hedge Trimming Shears

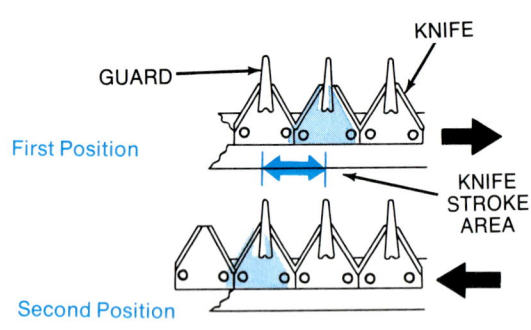

Fig. 3 — Knife Stroke Area

Fig. 4 — Types Of Cutter Bars

KINDS OF CUTTER BARS

In addition to the mower cutter bar shown in Fig. 1, there are cutter bars for grain combines, forage harvesters, and windrowers (Fig. 4).

Each of these machines has a cutter bar like one of those shown in Fig. 5. The cutter bars with lipped guards are used to cut grass and hay. Cutter bars with lipless guards are used to cut crops that can clog the cutter bar. For dense, matted crops, cutter bars with double knives are used. Special cutter bars are available with extra heavy, narrow spaced, or plateless guards.

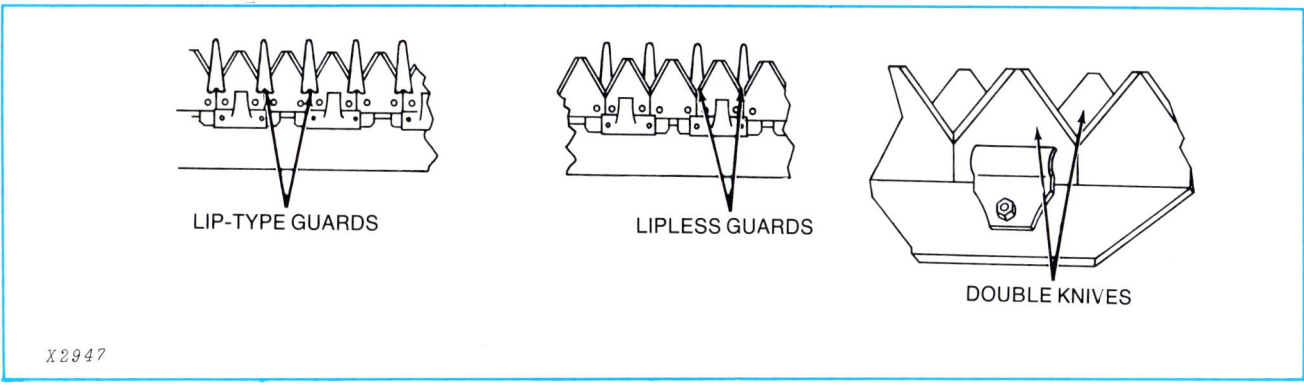

Fig. 5 — Three Types Of Cutter Bars

CUTTER BAR PARTS

Most cutter bars have the following basic parts (Fig. 6):

DRIVES

Read the "Special Drives" part of the FOS manual Power Trains to get a clear understanding of the details of cutter bar drives.

For the purposes of this book, there are only two kinds of cutter bar drives: Pitman drives and Pitmanless drives. Pitman drive knife speed is 1600 to 2000 strokes per minute. Pitmanless drive knife speeds are faster; 1800 to 2200 strokes per minute, and there is less vibration. The most common drive being manufactured is the Pitmanless drive.

BAR AND INNER AND OUTER SHOES

All the parts of a cutter are mounted on a bar (Fig. 6). There are gauge shoes on the inner and outer ends of the bar. Under each shoe there is a subsole that skids along over the ground and holds the cutter bar off the dirt in proper cutting position.

Self-propelled windrowers have adjustable gauge shoes, gauge wheels, or rollers on each end of the cutter bar.

KNIFE

The knife on a cutter bar has a head, a back, and sections.

The **knife head** is the connection point between the knife and the drive (Fig. 6).

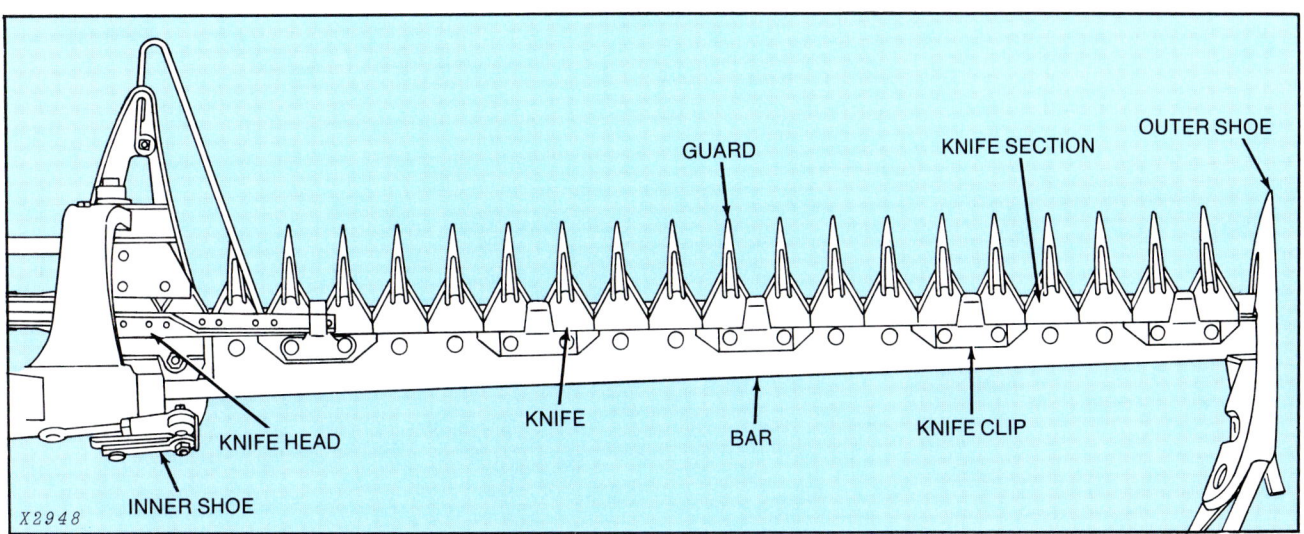

Fig. 6 — Mower Cutter Bar Parts

4 Mowing Cutter Bars

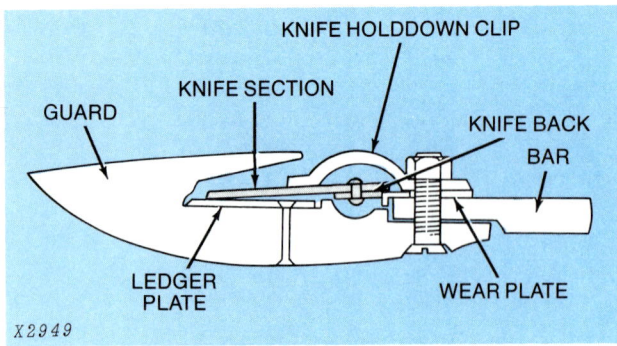

Fig. 7 — Relative Position Of Knife To Other Parts Of The Cutter Bar

The **knife back** is a flat, steel strip that the knife sections are riveted to (Fig. 7).

Knife sections are the individual cutting units on a cutter bar. There are four kinds (Fig. 8):

- **Smooth**
- **Top-serrated**
- **Bottom-serrated**
- **Armored**

SMOOTH SECTIONS are used to cut fine stemmed crops that buildup on the cutter. The smooth, chromed sections shed buildup better than plain steel sections.

TOP-SERRATED SECTIONS are used to cut coarse, stiff-stemmed crops such as straw, alfalfa, clover, and timothy. The serrations grab the crops so they don't push out in front of the sections. Top serrated sections also hold their cutting edges well; and if chromed they will shed crop buildup and juices well. Serrated sections are widely used on self-propelled units where the cutter bar is off the ground and away from fine dirt.

BOTTOM-SERRATED SECTIONS are used to cut the same crops as top serrated sections. However, with the serrations on the bottom, the sections can be sharpened.

ARMORED SECTIONS (not shown) are coated on the bottom with hard, tungsten carbide. They stay sharp longer than regular sections, and can be used anywhere serrated sections can. They can sometimes be used in place of smooth sections.

The guards (Fig. 9) protect the knife and act as a fixed, shearing edge for the moving section. Guards also divide the plants and guide them into the sections.

Different designs are available for special jobs. The kind of crop and conditions determine the proper guard.

GUARDS

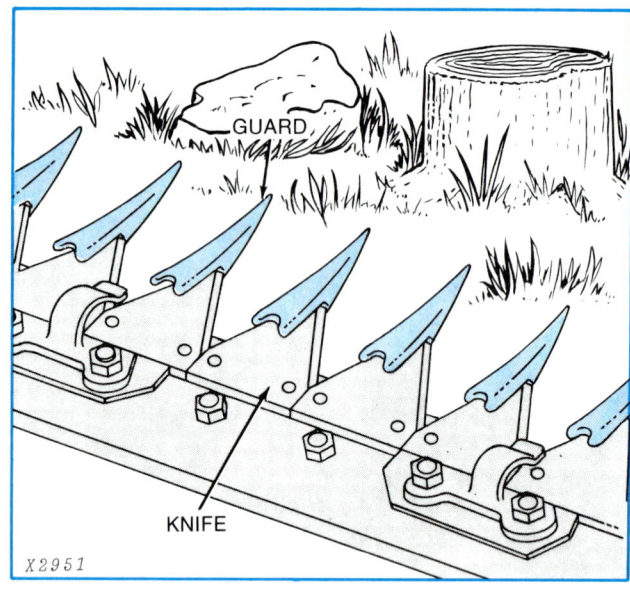

Fig. 9 — Guards Protect The Knife Sections

The **guards** (Fig. 9) protect the knife and act as a fixed shearing edge for the moving section. Guards also divide the plants and guide them into the sections for cutting.

Different designs are available for special jobs. The type of crop or conditions determine the proper guard.

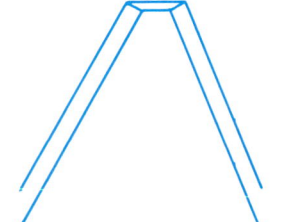 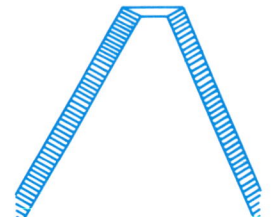 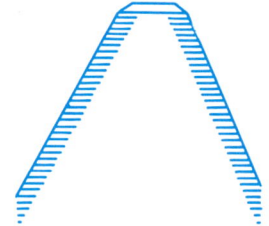

Fig. 8 — Three Kinds Of Knife Sections (Armored Not Shown)

Types Of Guards

The major types of guards are:

- **Rock guards**
- **Regular guards**
- **Lipless guards**
- **Two-tined guards**

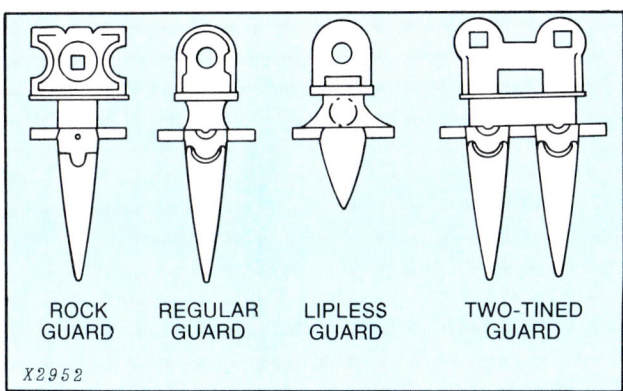

Fig. 10 — Major Types Of Guards

Rock Guards are the strongest guards made. They are made of steel or malleable iron, and have a 3-inch space from point to point.

Regular Guards are similar to rock guards in design and spacing. But, they are weaker and are only made of malleable iron.

Lipless Guards are special-purpose guards used where crop penetration is a severe problem. The section rides on top of the guard and cuts with less choking.

Two-Tined Guards are popular because they mount in pairs and are cheaper. They have a low profile and a slim design, giving good penetration. These guards are usually used on self-propelled machines but are rapidly gaining popularity on mower cutter bars, if stones are not a problem.

A **ledger plate** (Fig. 7) is the cutting part of a guard and can have either smooth or serrated edges. Ledger plates are clipped or rivited to the guard.

Two-tined guards do not have ledger plates. Instead, the surface where the section rides is machined to a sharp cutting edge.

WEAR PLATES AND KNIFE CLIPS

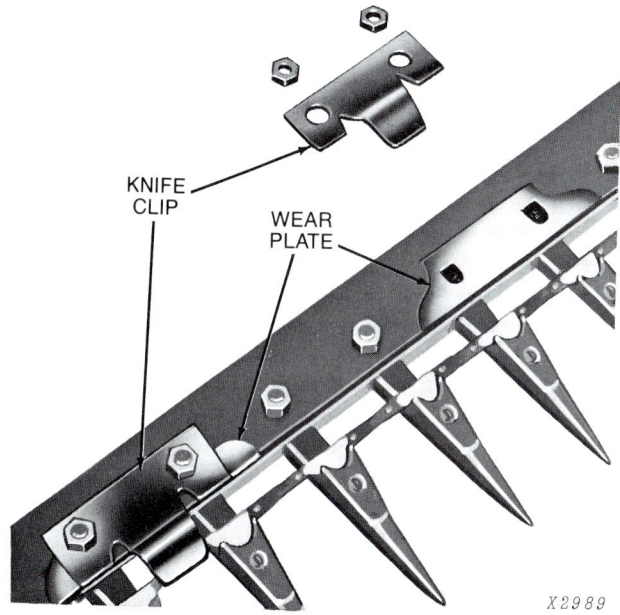

Fig. 11 — Wear Plates And Knife Clips

Wear plates (Fig. 11) guide the knife back. They keep the knife back positioned with the ledger plates to maintain shearing action. As they wear, they can be adjusted at their slotted mounting holes.

The **knife hold-down clips** (Fig. 11) hold the knife snugly against the ledger plates for clean cutting.

MAINTENANCE AND ADJUSTMENT

A cutter bar must be cared for and adjusted. Remember, no matter how well a cutter bar is designed, it will only operate as well as it is adjusted and maintained.

The keys to good cutter bar operation are:

1. **Maintenance**

2. **Adjustment**

6 Mowing Cutter Bars

MAINTENANCE OF THE CUTTER BAR

These components hold the knife and guard in place: wear plates, knife hold-down clips, knife-head guide plates, and drives.

The knife must be sharp. All the guards must be aligned with the knife. And, the wear plates, knife hold-down clips, and knife-head guides must be adjusted as they wear. Excessive play, from wear, will increase the wear rate.

On mowers, the lead angle of the cutter bar has to be correct or there will be excessive wear on the knife head. Also, for a good knife cut, the cutter bar must be set at a right angle to the tractor.

Poor maintenance of these parts can cause plugging, rob power, and slow down mowing.

If the drive mechanisms of a mower are not maintained at the recommended intervals, the bearings, belts, chains, or gears may fail at a critical time.

ADJUSTMENT OF THE CUTTER BAR

Adjustment of the cutter bar usually involves more than one step; making one adjustment may require another. For example, when wear plates require adjustment, knife hold-down clips and knife-head guides will probably need checking too.

Complete cutter bar adjustment includes the following jobs:

1. **Adjusting the Knife:**
   a. Register
   b. Head Guides
   c. Wear Plates
   d. Hold-Down Clips

2. **Adjusting the Guards**

3. **Adjusting the Cutter Bar Lead**

4. **Adjusting the Operating Height:**
   a. Shoes
   b. Flotation

5. **Adjusting the Cutter Bar Tilt**

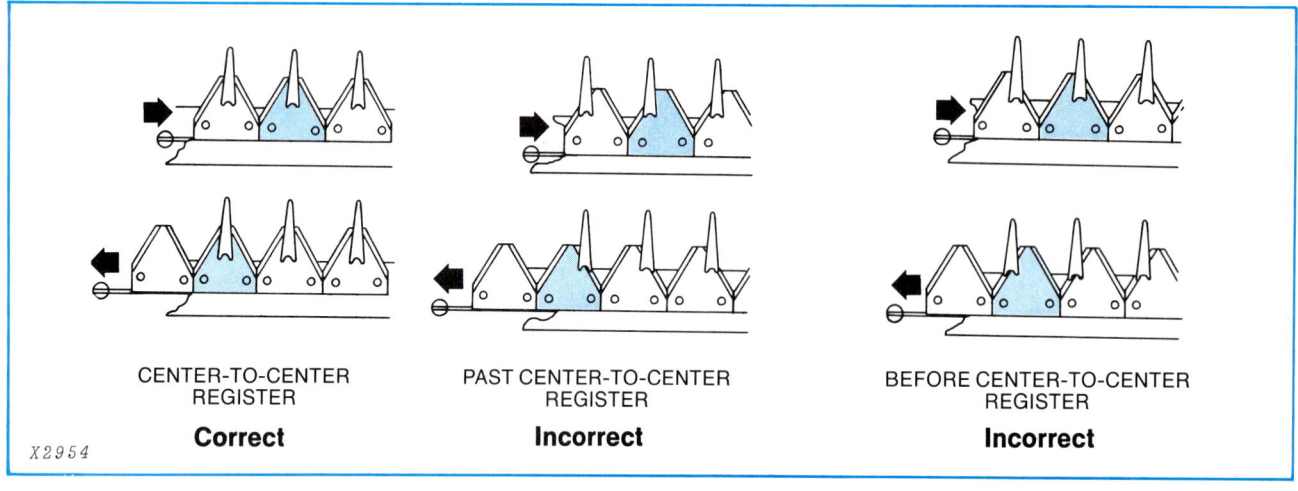

Fig. 12 — Three Types Of Knife Register

FOS — 56   Litho in U.S.A.

ADJUSTING THE KNIFE

A sharp, properly adjusted knife is the key to efficient mowing. The knife must run smoothly in the cutter bar. You should be able to move it easily by hand. Each section should be sharp and riveted tightly to the knife back.

Checking Knife Register

Knife register is an important adjustment on pitman drive cutter bars. Normally it is changed when the pitman is damaged or replaced, or if the cutter bar lead is changed.

Most pitmanless drive cutter bars, like balanced head and wobble knife drives, remain in register automatically because the reciprocating mechanisms are mounted on the cutter bar. The stroke remains the same, so the register does not change.

To check the register, be sure the drive mechanism is attached to the knife and the cutter bar is level and in cutting position. Rotate the drive by hand, and observe the position of the knife sections in relation to the guards at the inner and outer ends of the stroke.

The knife is in register if the knife sections are the *same distance* from the center line of the guards at the end of each stroke (Fig. 12).

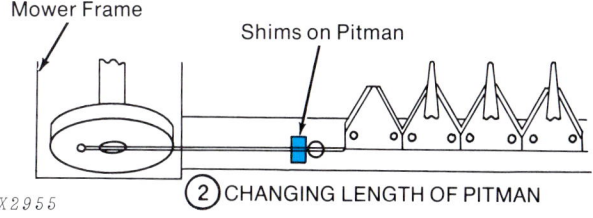

Fig. 13 — Two Methods Of Adjusting Knife Register

Most newer mowers don't have a knife register adjustment. But, on older mowers, adjust knife register by either moving the cutter bar in or out, or changing the length of the pitman (Fig. 13).

Adjusting Knife Head Guides

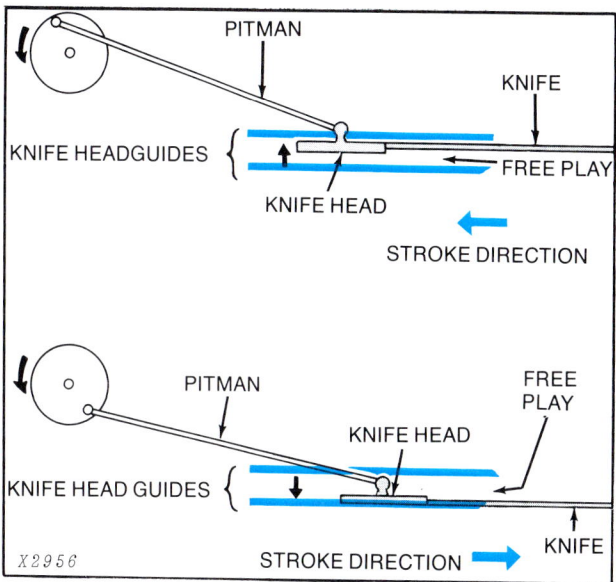

Fig. 14 — Excessive Vertical Play

When out of adjustment, the *knife head* will slap or hammer because of excessive vertical play (Fig. 14).

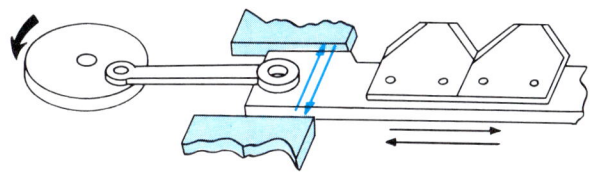

Fig. 15 — Excessive Side Play

Some pitmanless drives have side play and must be adjusted to compensate for wear (Fig. 15).

To check for wear on the knife head guides, grasp the knife head and move it up and down and side to side. If there is an unusual amount of play, adjust the knife head guide. Most knife head guides are adjusted by adding or removing shims (Fig. 16). The amount of vertical clearance is stated in the machine operator's manual. Use a thickness gauge to measure the spacing.

8 Mowing Cutter Bars

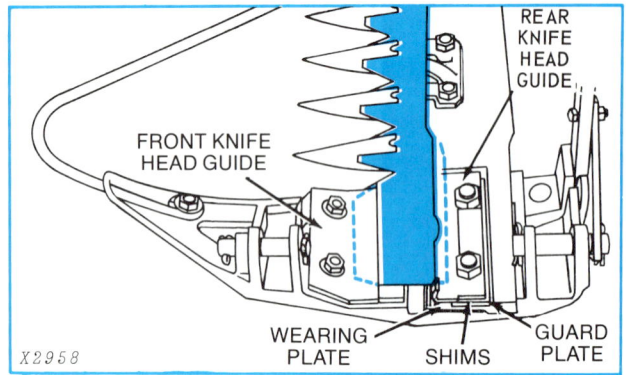

Fig. 16 — Knife Head Guide Adjustment

Adjusting Wear Plates

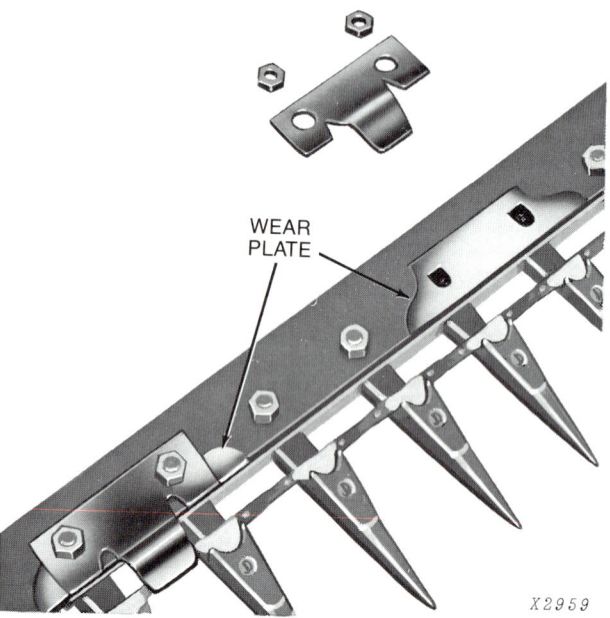

Fig. 17 — Wear Plate And Knife

Wear plates act as guides for the knife back (Fig. 17). They hold the knife back so the knives shear between the ledger plates and the knife sections.

Excessive play is caused by too much wear between the knife back and the wear plates. It will allow more vibration and cause fast wear.

As the wear occurs, move the wear plates forward, but *do not move them in too close*. Slots in the plates allow this adjustment.

Adjusting Hold-Down Clips

Knife hold-down clips hold the knife close to the ledger plate (Fig. 17). *If the hold-down clips hold the knife too loose*, ragged cutting and choking will result. *If they are too tight*, the knife will bind.

Most manufacturers prefer the knife tilted slightly so there is minimum clearance between the front of the ledger plate and the knife section tips. This insures a positive contact between the ledger plate and the knife section at the shear point. It gives a more positive cutting action and reduces draft. This calls for approximately a $\frac{1}{32}$-inch (1 mm) clearance between the rear of the ledger plate and the rear of the knife section. Others have no clearance. The knife lies flat on the ledger plate.

There are two ways of adjusting hold-down clips:

(1) Add or remove spacers to increase or reduce clearance between the clip and the knife section.

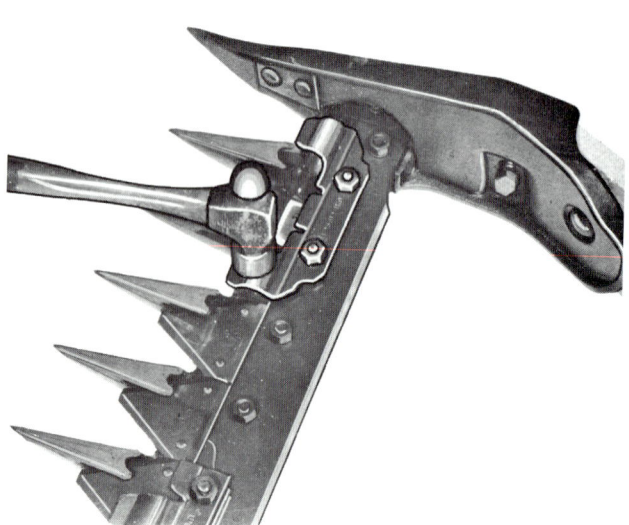

Fig. 18 — Setting Knife Clip Down

(2) The second method involves clips that are not mounted with spacers and must be adjusted with a hammer. To reduce clearance, strike the clip on the top with a two-pound (1 kg) hammer to bend it downward (Fig. 18).

Mowing Cutter Bars

them if they are dull or broken. Tighten all the plates to maintain the opposing shearing force to the cutting edges of the knife sections.

Aligning The Guards

Look for guards that are too high. Do not attempt to adjust the majority of the guards upward to align with a few that are too high.

When aligning the points of the guards, be sure the cutter bar is held rigid by the platform frame on windrowers or the inner shoe and drive unit on mowers.

Guards are made of cast, malleable iron or forged steel. Forged steel guards are less likely to nick, bend, or dull.

Straighten and realign guards made of either metal with a hammer, and smooth them with a file.

Fig. 19 — Increasing Clearance on Knife Clip

To increase clearance, strike clip on flat surface at the rear of the clip (Fig. 19).

ADJUSTING THE GUARDS

The guards protect knife sections and hold the ledger plates. They also act as dividers that comb through the crop separating the plants and guiding them into the knife. Guard points should be sharp, smooth, and aligned. Guard maintenance can be divided into three areas:

1. *Checking the condition of the guards.*

2. *Aligning the guards.*

3. *Adjusting the guard lip clearance.*

Checking The Condition Of Guards

Checking the condition of the guards can be done while they are mounted on the cutter bar. Look for wear, nicks, breaks, rust, and bending. Any of these things can cause improper operation. Use a file to remove rust and nicks, and to sharpen the guard points. If this cannot be done, replace the damaged guard, inspect the ledger plates, and replace

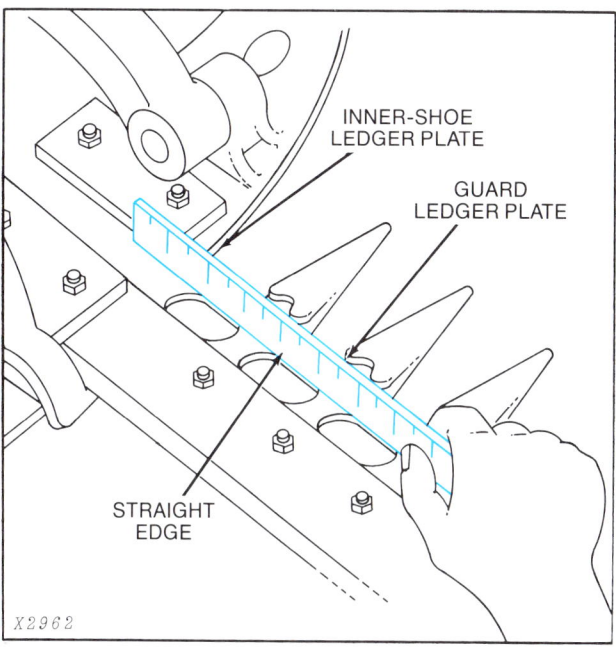

Fig. 20 — Checking Guard Alignment (Mower Shown)

For mowers; line up the first guard and the inner shoe ledger plate with a straight edge (Fig. 20).

If the guard needs to be raised, bend it up with a hammer. Strike the guard in the thickest place, usually just ahead of the ledger plate.

Disregard the position of the guard points; the guard ledger plates are what must be in line.

10 Mowing Cutter Bars

SETTING GUARD DOWN   SETTING GUARD UP

Fig. 21 — Setting Guards Down (Mower Shown)

If the guard must be lowered, hold a heavy hammer under the base of the guard to help steady the blows (Fig. 21). Pound in front of the ledger plate. Be careful not to bend the guard lip.

Perform the same procedure at the outer shoe. After the guards have been aligned at both ends, sight down the cutter bar and bend any of the other guards which are out of adjustment.

NOTE: *Special guard adjusting tools are also available for aligning guards.*

Aligning the guards on a combine, forage harvester, or windrower is slightly different. There is no inner or outer shoe as on mowers. Therefore, the starting point is at the driven end of the knife. Align the first two or three guards to allow the knife head minimum friction or binding on the guards as it moves back and forth. Then align the rest of the guards by sighting down the length of the cutter bar. On long knives it is virtually impossible to get it absolutely straight. A slight, gradual "droop" will do no harm.

Adjusting Guard Lip Clearance

Adjusting the guard lip clearance is the final step in guard maintenance. Sometimes aligning the guards will change the guard lip clearance; this is the reason for doing this adjustment last. Hold the clearance between the rear of the guard lip and the knife at approximately ⅜ to ½-inch (10 to 13 mm) (Fig. 22). Too small a clearance will create plugging and binding of the material because it cannot pass through after it has been cut. Too much clearance will leave the knife unprotected. (On guards with no ledger plate, the clearance at the back of the lip should be $1/32$ to ⅛ inch (1 to 3 mm) more than at the front.)

FOS — 56   Litho in U.S.A.

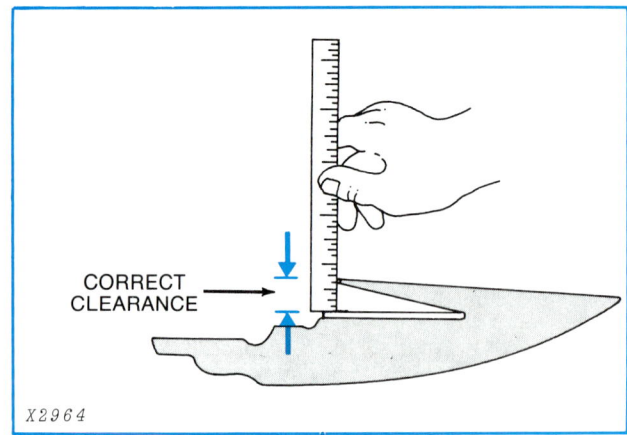

Fig. 22 — Checking Guard for Lip Clearance

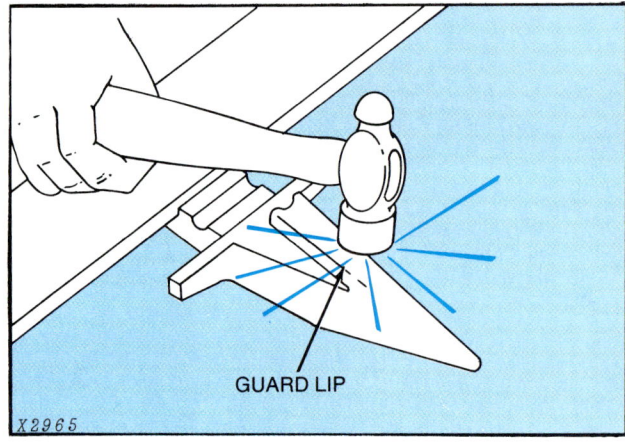

Fig. 23 — Reducing Guard Lip Clearance

To decrease clearance, tap lightly on the lip with a hammer (Fig. 23).

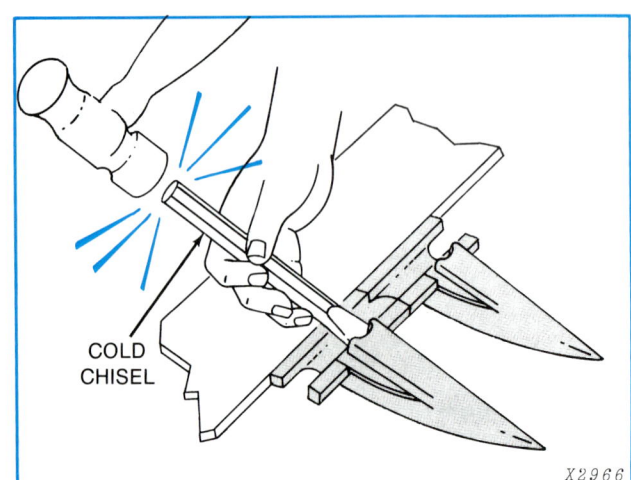

Fig. 24 — Increasing Guard Lip Clearance

To increase clearance, use a chisel as a wedge (Fig. 24). Be careful! Overbending can break the guard lip.

ADJUSTING CUTTER BAR LEAD (MOWERS ONLY)

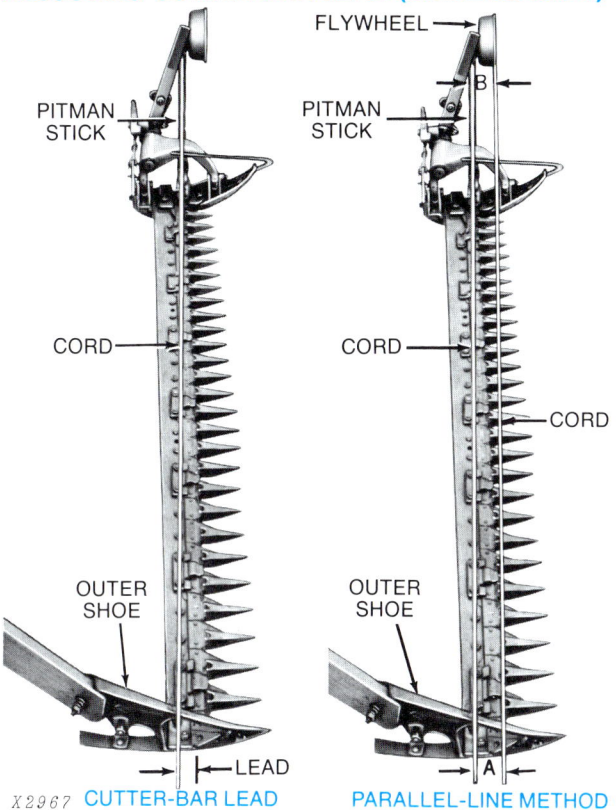

Fig. 25 — Cutter Bar Lead

Cutter bar lead applies only to cutter bar **mowers.** Combines, forage harvesters, and windrowers do not have this adjustment because the platform stabilizes the cutter bar and will not allow one end to lag behind the other while cutting.

However, mowers have only one end of the cutter bar connected to a solid object; the bar is pushed back by plants when it is cutting. For this reason, the outer end should be ahead of the inner end of the cutter bar (Fig. 25) when the mower is not in operation. This is called **cutter bar lead.** Cutter bar lead is divided into **primary lead** and **secondary lead.**

Primary Lead

When mowing, the pressure of the plants against the cutter bar pushes the outer end back until the cutter bar is at right angles to the forward motion. This is called *primary lead.*

Secondary Lead

When operating, the pitman must be at right angles to the line of travel of the mower. This is called *secondary lead.*

Adjustment

Normal wear will cause the cutter bar lead to decrease until it needs adjustment. To check and adjust, proceed as follows:

1. Park tractor and mower on a level surface.
2. Be sure cutter bar is flat on the ground.
3. Pull outer end of cutter bar backward by hand to take up any slack.
4. Find a guideline from which you can measure the lead (Fig. 25).

The guideline should be at a right angle to the forward motion of the mower and parallel to the pitman on pitman mowers.

One of two basic methods of establishing a guideline is called the **parallel-line method** (Fig. 25). Stretch a cord from the pitman bearing across the cutter bar while making sure the cord is parallel to the pitman. Secure the cord at both ends.

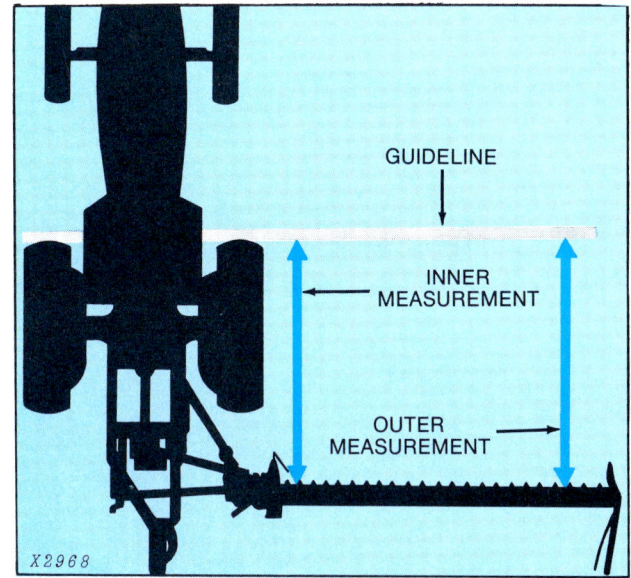

Fig. 26 — Straight-Board Method

The second is called the **straight-board method.** Place a straight board in front of the rear tires of the tractor, as shown in Fig. 26. Secure the board so it cannot be moved. Measure from the board guideline back to the knife section rivets.

The distance from the outer end of the cutter bar should be *shorter* than from the inner end, measured from the guideline. The amount varies depending on the length of the cutter bar and the manufacturer's recommendation. The normal range is ¼-inch per foot (6 mm per 305 mm) of cutter bar. The longer the cutter bar, the more lead is necessary.

12 Mowing Cutter Bars

Next, adjust the cutter bar primary and secondary lead. Three methods are most common:

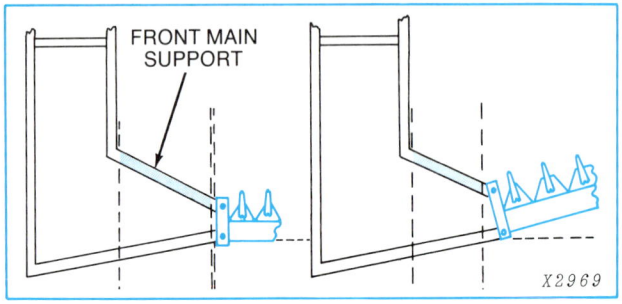

Fig. 27 — Shortening Front Support

1. Shorten the front main cutter bar support (Fig. 27).

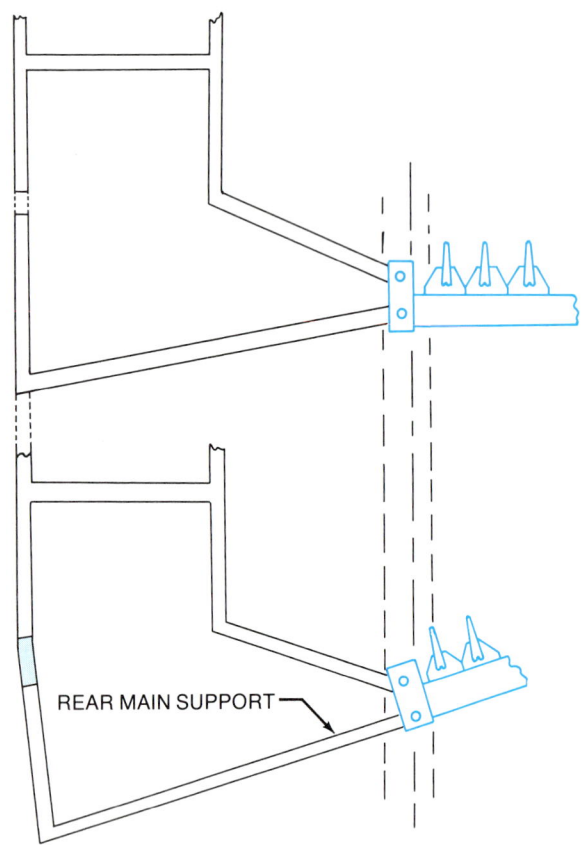

Fig. 28 — Lengthening Rear Support

2. Lengthen the rear main cutter bar support (Fig. 28).

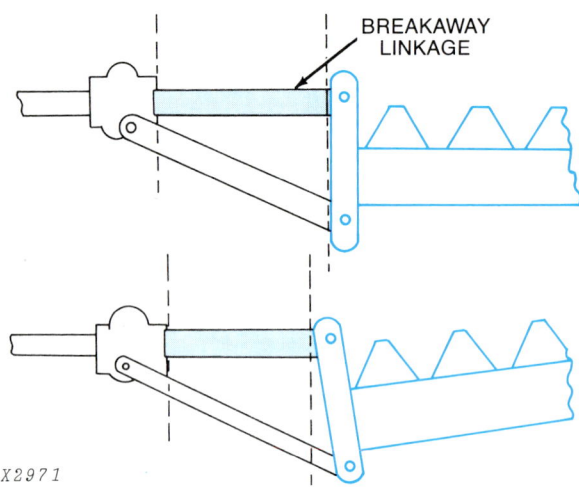

Fig. 29 — Shortening Length Of Breakaway Linkage

3. Shorten the length between breakaway linkage and clamps (Fig. 29).

Select the method used for your mower and make the necessary adjustment.

ADJUSTING THE OPERATING HEIGHT

Height of cut depends on the type of crop and ground conditions. In stony conditions, a longer length of cut would be best to minimize cutter bar damage. If the ground is smooth and rock-free, a shorter length of cut is possible.

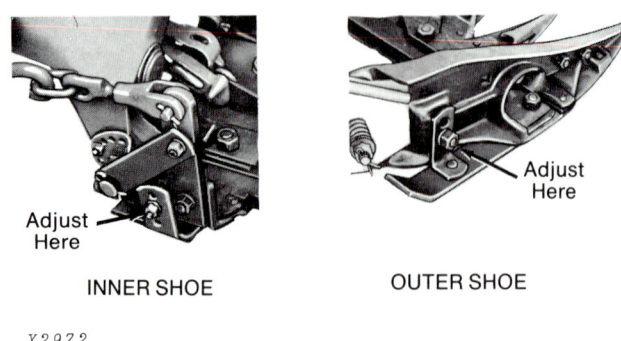

Fig. 30 — Mowing Height Adjustment For Mower

Mowers have adjustable runners on their shoes (Fig. 30) which provide the height adjustment. Move the bolts to new holes to set the cutter bar height.

Windrowers have adjustable shoes, rollers, or wheels to set cutter bar height (Fig. 31). The shoes on the platform of a windrower are adjusted in the same manner as the runners on a mower.

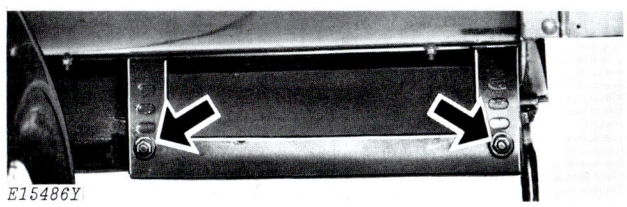

Fig. 31 — Windrower Gauge Shoe

Fig. 32 — Gauge Roller

ed so the guard points can penetrate and lift the crop for proper cutting.

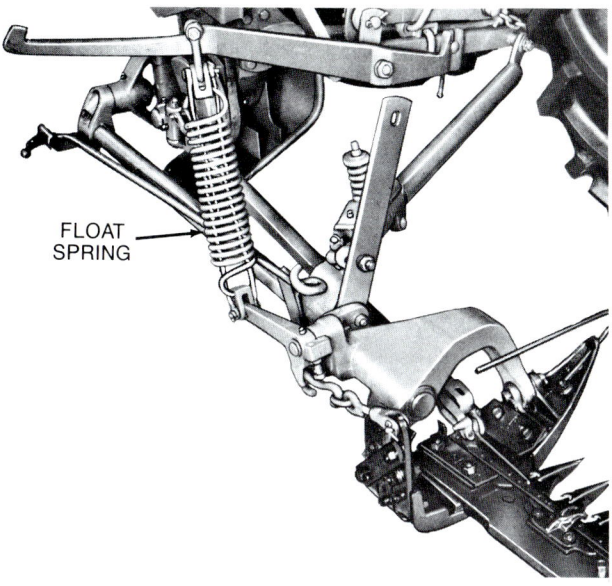

Fig. 33 — Cutter Bar Float Spring

Gauge wheels or rollers (Fig. 32) replace the shoes on some platforms but are adjusted in the same way.

Closely related to height of cut is cutter bar *flotation*. The cutter bar float spring (Fig. 33) should be adjusted so that the cutter bar will follow the ground contour so it can do a good job of cutting at all times. Too much flotation will cause the bar to bounce, leaving good crops uncut. Too little flotation will allow the bar to ride too heavily on the ground, causing excessive guard and knife damage. For exact adjustment, see the operator's manual.

ADJUSTING THE CUTTER BAR TILT

Cutter bar tilt is vital to a good job of cutting. **Tilt** refers to the angle of the cutter bar in relation to the ground. This tilt is adjustable by means of a lever (Fig. 34) or other device. The ideal angle is parallel to the ground but often this is impractical due to crop or ground conditions.

Mowers usually have an 8 to 10 degree upward and 8 to 10 degree downward tilt range. When the crop is down or tangled, a downward tilt is recommend-

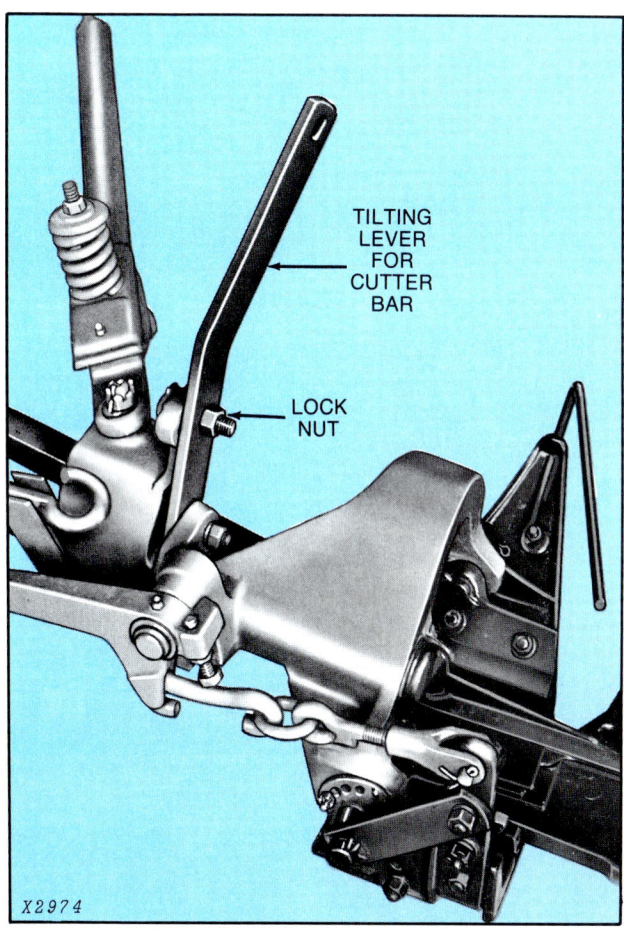

Fig. 34 — Cutter Bar Tilt Adjustment

14 Mowing Cutter Bars

Tilting the cutter bar downward increases the possibility of guard and knife damage because they are closer to the ground. In stony, rough conditions upward tilting is recommended because it raises the guard points and reduces damage to the cutter bar.

Windrowers and combines have a different cutter bar adjustment, but the tilt principle remains the same. It is difficult to change the cutter bar angle because it is an integral part of the platform. Therefore, the only way the tilt can be changed is by moving the entire platform. Most windrowers have a downward tilt built into the cutter bar so they can cut the crop short in all conditions.

If adjustment will not correct the problems, a defective part may be the problem. The defective part must be repaired or replaced.

REPAIR AND REPLACEMENT

The decision when and how to replace parts can be a problem unless you have a good knowledge of cutter bar repair. This section will explain how to make repairs and when to replace worn or damaged parts.

SERVICING THE KNIFE

The knife sections must be sharp; if not, they should be sharpened or replaced. Damaged sections must also be replaced. In addition, the knife back must be straight; if not, excessive wear on parts or binding will result.

The knife must be removed from the cutter bar for service. Some repair or replacement can be made while the knife is in the cutter bar, but it is not recommended.

REMOVING THE KNIFE

Removing the knife is relatively easy. But, be careful at all times. You can easily lose a finger by careless handling of cutter bar knives. For easy handling, oil the knife and run it back and forth for a minute to loosen any rust deposits. If the knife is gummed up, spray water over it while the knife is running.

Then block the cutter bar up with wooden blocks for easier access to the knife (Fig. 35).

Fig. 35 — Blocking Up The Cutter Bar

Next, remove the connector from the knife head. Three types of connectors are common: pin, bolt, and spring-loaded.

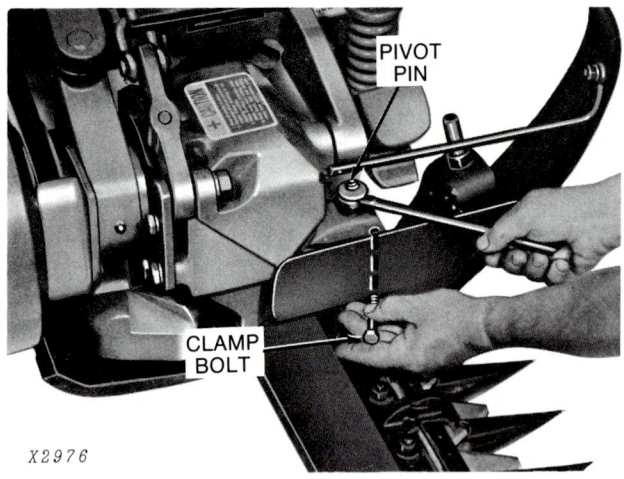

Fig. 36 — Pin Connector

On *pin connectors*, remove the clamp cap screws and pull the pin from the knife head (Fig. 36).

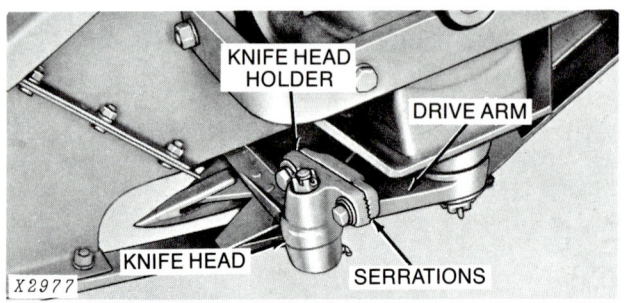

Fig. 37 — Bolt Connector

On *bolt-types*, remove the cap screws that attach the knife-head holder (Fig. 37) to the drive arm.

FOS — 56   Litho in U.S.A.

Mowing Cutter Bars 15

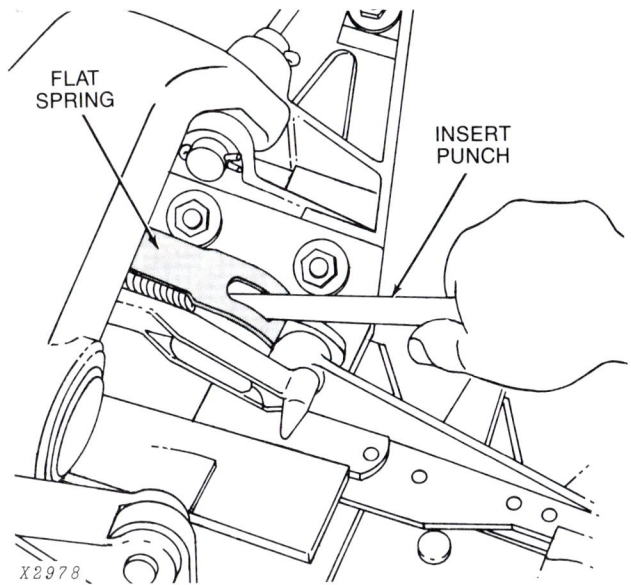

Fig. 38 — Spring Loaded Connector

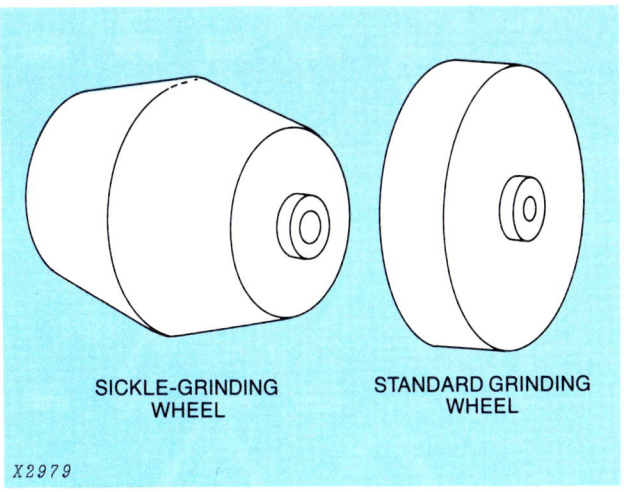

Fig. 39 — Grinding Wheels

On *spring-loaded types,* insert a punch in the hole of the flat spring (Fig. 38), force the plunger back, and then push the flat spring down between the pitman straps to free the knife head.

Pull the knife out by hand.

CAUTION: If the sections hang up on the ledger plates, maneuver them with a wooden block, — not your fingers.

Grasp the knife on its back side, away from the knife sections, with *both hands.* This will keep your fingers out of a vulnerable position.

REPAIRING THE KNIFE

Straightening the knife is easy if it is not bent severely. Remove the knife as described above and lay it on a smooth, flat surface. Sight along the edges to determine where it is bent. If it is crooked, clamp the knife in a vise and straighten it by careful bending.

Knife sections can be *sharpened* with good results if done properly. Overserrated sections cannot be sharpened satisfactorily; they must be replaced when dull or broken. Take special care while grinding sections.

For best results, use a special sickle-grinding wheel (Fig. 39) of medium coarse 30 to 60 grit size. It will grind the original bevel on the sections.

A standard grinding wheel can be used, but it is very easy to grind the wrong bevel onto the sections. The same holds true for a hand file.

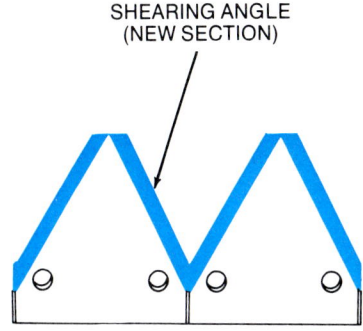

Fig. 40 — Original Shear Angle And Bevel

Keep the shearing angle of the section the same as the original edge appeared. The face of the bevel should be approximately ¼-inch (6 mm) wide (Fig. 40).

When sections have been ground EXTENSIVELY (Fig. 41) replace them because they no longer have a heat-treated surface, and wear occurs very rapidly. **Do not overheat the section when grinding.** Most sections have a hardened edge to help hold a sharp edge. Repeated or improper sharpening will grind away the hardened edge and rapid wear will occur.

16 Mowing Cutter Bars

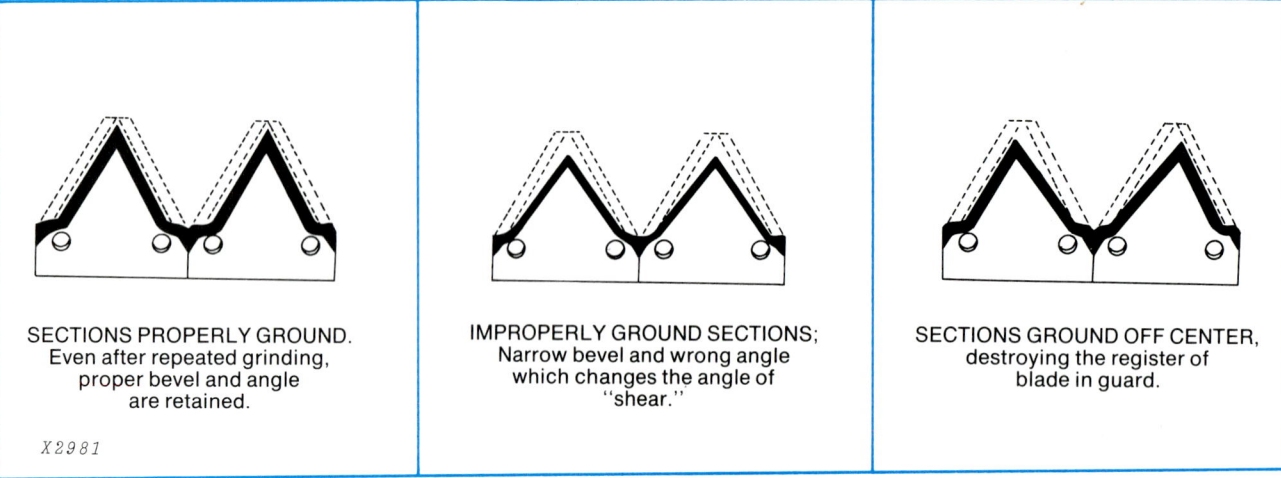

Fig. 41 — Knife Sections That Have Been Sharpened Several Times

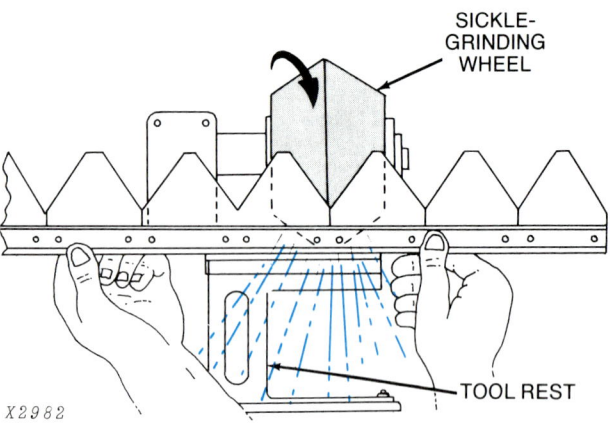

Fig. 42 — Correct Grinding Method

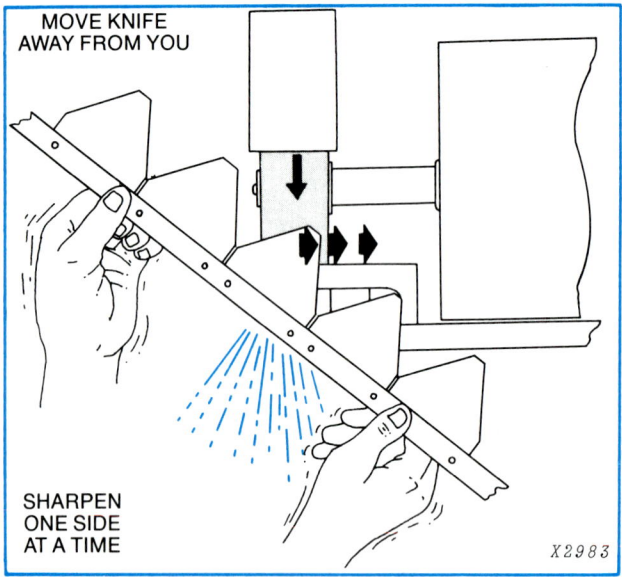

Fig. 43 — Using A Standard Grinding Wheel

If you are using a beveled sickle grinding wheel, hold the knife assembly in both hands and place it lightly on top of the wheel (Fig. 42). The knife back should rest on a stable, flat surface for easy control. The grinding wheel should turn against the edges of the sections so feathering will not occur.

If you must use a standard grinding wheel, sharpen one side of the section at a time (Fig. 43). Use only the curved face of the wheel and not the sides. Move the knife away from you as you sharpen.

If the section is broken, replace it. It is possible to replace sections when the knife is in the cutter bar, but a section clamp must be used. The old rivets can be sheared off with a chisel and hammer and the new section and rivets can be mounted using the clamp. The clamp forms the rivet head.

This method of replacing sections should be used only if a few sections need replacing. When all the sections are to be changed, knife removal is essential and a knife repair block or a vise should be used.

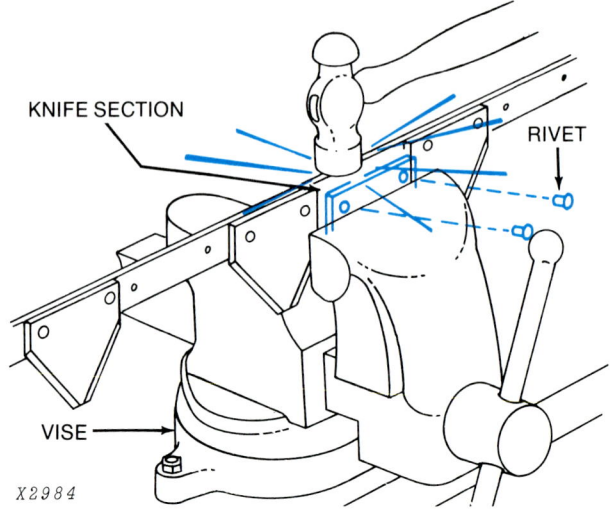

Fig. 44 — Shearing Rivets In A Vise

If you use a vise, place the knife tips down between the jaws with the knife back resting on the stationary jaw. The knife should be held loosely so that the section can be driven between the jaws with a two-pound (1 kg) hammer to shear the rivets as shown in Fig. 44.

Replace the sections alternately so that the new sections remain in proper relationship. After every other section has been replaced, go back and change the rest. This keeps the knife straight and eliminates shifting of rivets when you tighten the sections to the knife back.

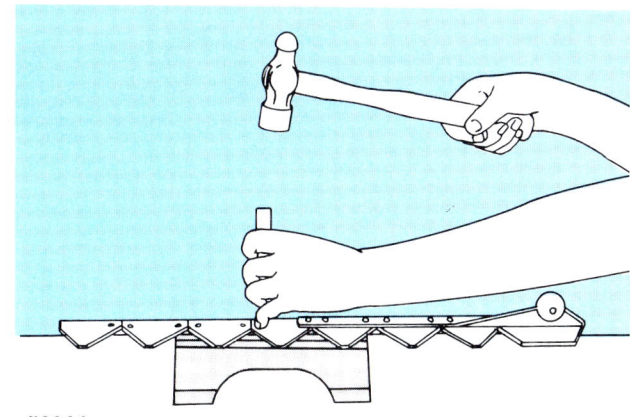

Fig. 46 — Making Rivet Head

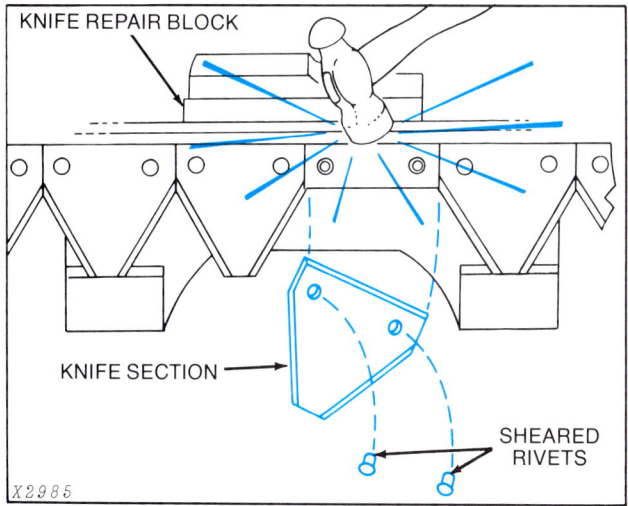

Fig. 45 — Shearing Rivets In A Repair Block

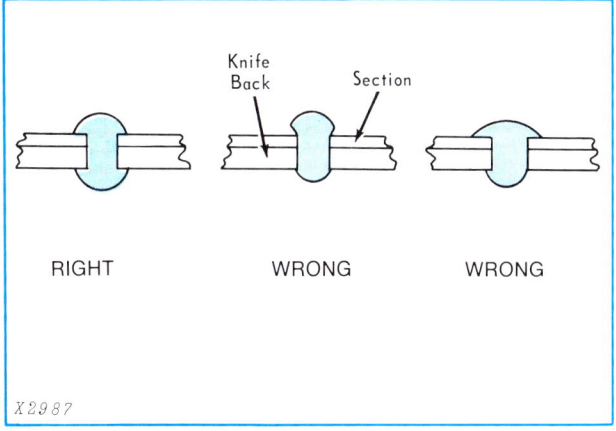

Fig. 47 — Correct and Incorrect Ways To Form Rivets

If you use a knife repair block, place the knife tips down, in the slot provided (Fig. 45), and replace the sections in alternate order.

Strike the back edge of the section with a 2-pound (1 kg) hammer and shear the rivets off. If the rivets are punched out, there is a good chance of enlarging the holes in the knife back.

To install new sections, you must select the correct rivet size. A rivet should extend 1½ to two times its diameter through the knife section.

Insert two rivets through the knife back and section and place the knife on a riveting block with the bevel side of the section up. Strike the rivet with a rivet setting tool, using firm, direct blows (Fig. 46). This will expand the rivet and make a good head.

A rivet that is not formed properly will quickly loosen and fail (Fig. 47).

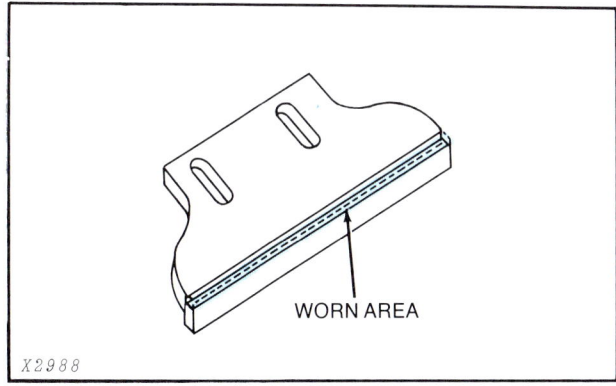

Fig. 48 — Wear Plate Wear Area

Wear plates cannot be repaired. They must be replaced. When the plate wears to one-half its thickness, replace it (Fig. 48).

18 Mowing Cutter Bars

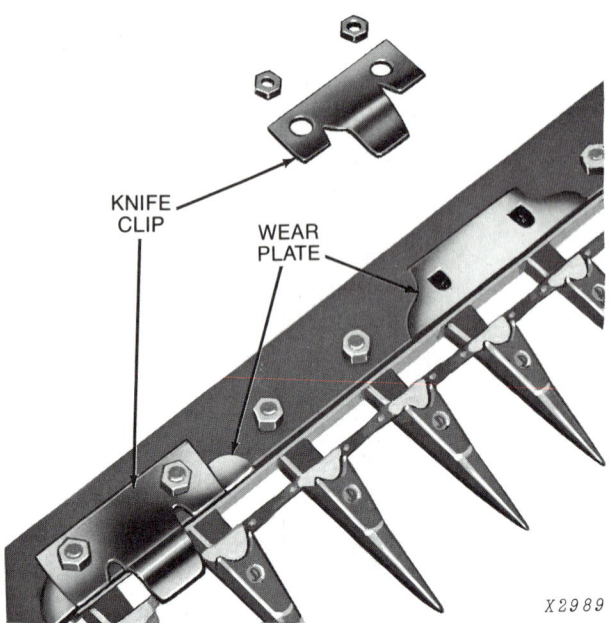

Fig. 49 — Replacing Wear Plate And Knife Clip

Remove the guard nuts, knife hold-down clip, and old wear plate (Fig. 49), and replace them with a new plate. Adjust the wear plate so the knife is free moving through one complete cycle. Tighten the guard nuts.

Replace the *knife hold-down clips* if they cannot be properly adjusted. To remove the clip, remove the guard nuts and change the clip. Tighten the guard nuts.

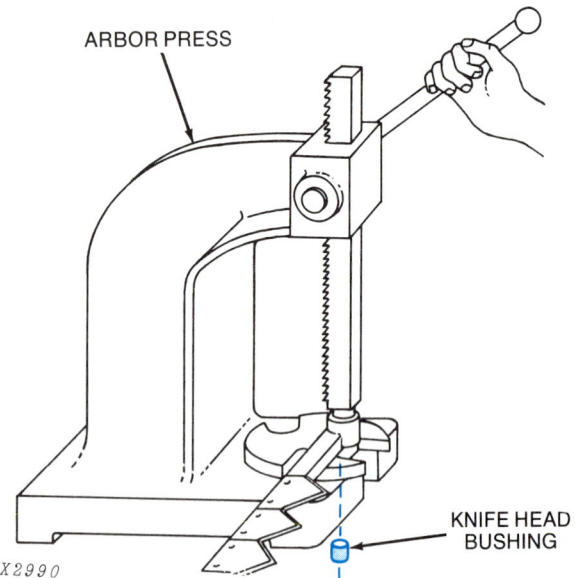

Fig. 50 — Removing A Knife Head Bushing With An Arbor Press

The *knife head* is attached to the drive unit by a pin or shaft. Replace the knife head bushing if it becomes worn. A worn bushing causes excessive vibration and additional wear on other cutter bar parts. The simplest way to check for a worn bushing is to check the play between the pin and bushing.

To replace the bushing, use a press or a punch (Fig. 50). If you use an arbor press, place a pipe or steel rod, with the same diameter as the bushing, over the bushing and press it out. Press the new bushing into the knife head in the same manner.

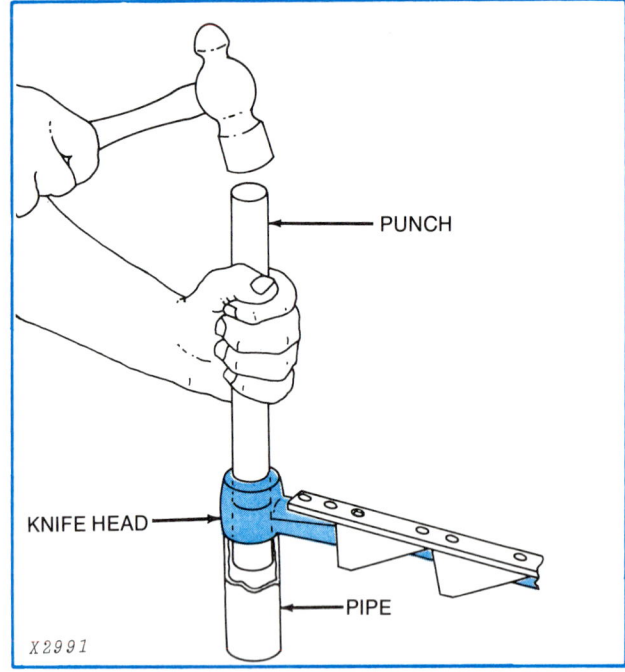

Fig. 51 — Removing A Knife Head Bushing With A Punch

If you use a punch, position the knife head over the end of a pipe and tap the punch with a hammer to remove the bushing (Fig. 51). Be sure the punch has the same diameter as the bushing. Replace the new bushing in the same manner. Check the pin for wear and replace it if required.

SERVICING THE GUARDS

Once a cutter bar *guard* is broken, it is almost impossible to repair. It must be replaced.

To replace a guard, remove the guard bolt, which may also secure the wear plate and knife hold-down clip (Fig. 52).

If the guard is not broken but is badly bent and nicked, try to straighten the guard. Place a pipe over the point and bend it to its original position. File the nicks.

Mowing Cutter Bars 19

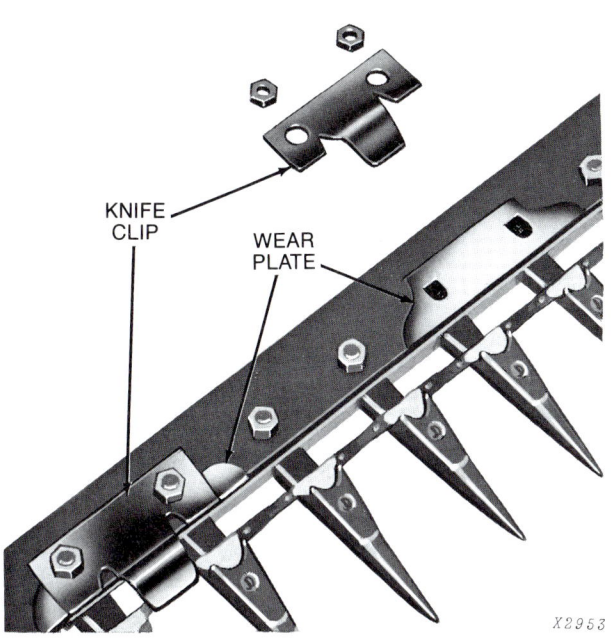

Fig. 52 - Guard Replacement

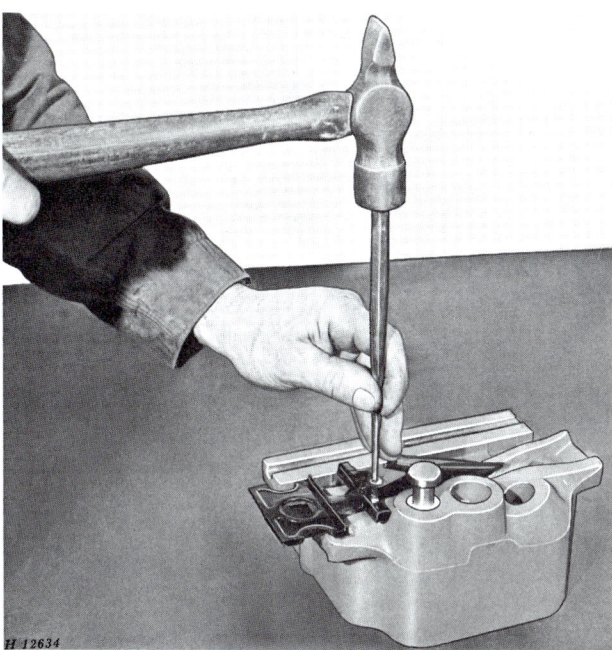

Fig. 54 — Removing Riveted Ledger Plate

The *ledger plates* must be kept in good condition. Poorly maintained ledger plates cause ragged cutting and side draft. Since most ledger plates are serrated, they cannot be sharpened, only replaced. The first thing to determine is how they are held in place.

Quick-attach ledger plates are held in position by pins, springs, or wedging. On those using pins, replace the plate by driving the pin up from the bottom of the guard then driving it back into position to hold the new plate. Plates held by springs merely require snapping out the old plate and putting in the new one.

Ledger plates that are riveted may be replaced without removing the guard from the cutter bar, but it is easier if the guard is removed.

Place the guard on a solid surface with a recess in the center and drive the rivet out from the top (Fig. 54). A guard repair block comes in handy for this job.

If you cannot drive the rivet out, you may be forced to drill it out. If it is worn away, countersink the rivet hole on the lower side (face) of the guard with a countersink or a $5/16$-inch (8 mm) twist drill. This will allow you to smooth the bottom of the guard and have no rivet head protruding.

Place the new plate in position and insert the rivet from the top.

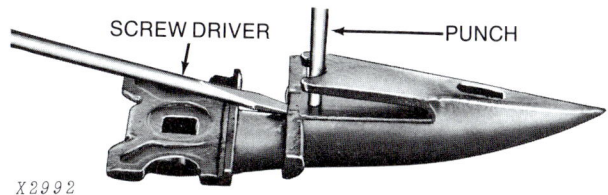

Fig. 53 — Removing Ledger Plate

Remove wedged in plates with a screwdriver (Fig. 53). Raise the rear edge of the plate and pry against the lip of the guard to slide it out.

To put in a new plate, drive it forward until it drops into position. Then set the plate by tapping around the rear of the plate.

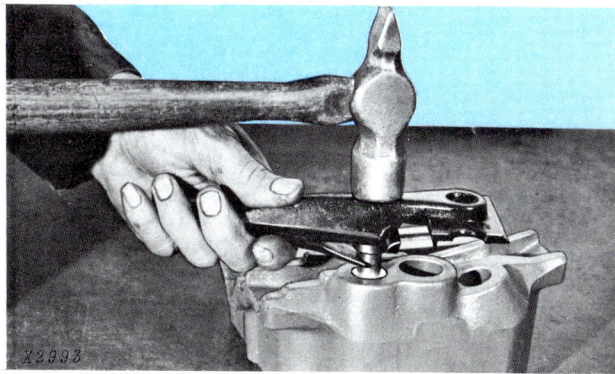

Fig. 55 — Forming Rivet Head

Form rivet head on the bottom of the guard with a hammer by holding the guard on a solid surface, upside down (Fig. 55).

FOS — 56   Litho in U.S.A.

File both the top and bottom of the rivet until it is even with the plate and guard.

SERVICING THE SHOES

Some mowers have inner and outer shoes which have ledger plates. When these plates become worn or damaged, plugging will occur.

Replacing Inner Shoe Plate

To replace the inner shoe plate, remove the shoe from the cutter bar. Remove parts that will interfere with removal of the ledger plate.

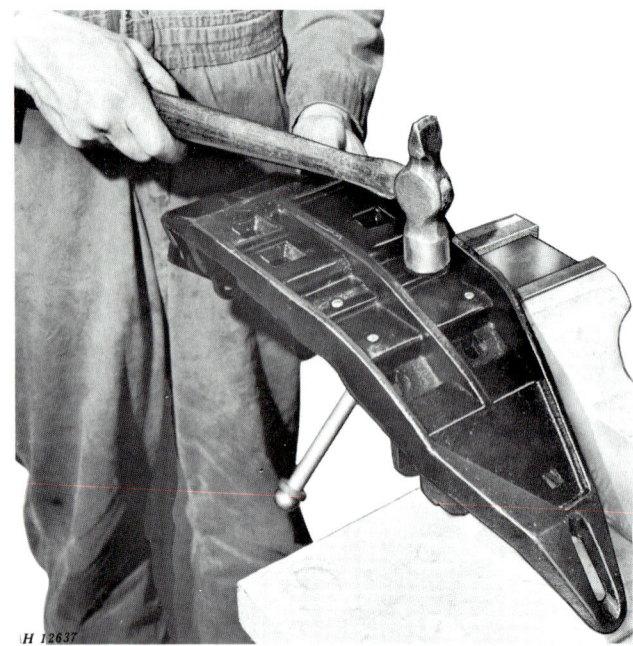

Fig. 57 — Replacing Ledger Plate On Inner Shoe

Replacing Outer Shoe Plate

To replace the outer shoe plate, remove the shoe from the cutter bar. Remove parts that will interfere with removal of the ledger plate.

Fig. 56 — Removing Rivets From Inner Shoe

Place the inner shoe in a vise and drive out the rivets holding the plate (Fig. 56).

Remove the shoe from the vise and place a piece of steel ½-inch thick, 2-inches wide, and about 8-inches long (13 mm thick, 50 mm wide, and about 200 mm long) in the vise. This piece will be used as a backup bar.

Position a new plate on the shoe with new countersunk rivets. Turn the shoe over, face down, so that the backup bar is under the ledger plate. Peen over the rivets on the bottom of the shoe (Fig. 57).

Replace all parts removed from the shoe and reinstall it on the cutter bar.

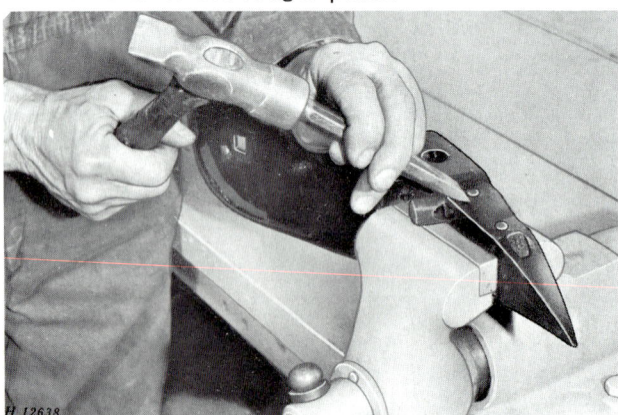

Fig. 58 — Removing Rivets From Outer Shoe

Place the inner shoe in a vise, upside down, and chisel the head off the rivets (Fig. 58). Punch out the rivets and remove the old plate. Remove the shoe from the vise and place a piece of steel ½-inch thick, 2-inches wide, and about 8-inches long (13 mm thick, 50 mm wide, and about 200 mm long) in the vise. This piece will be used as a backup bar.

Position a new plate on the shoe with new countersunk head rivets. Turn the shoe over, face down, so that the backup bar is under the ledger plate. Peen over the rivets on the bottom of the shoe (Fig. 59).

Replace all parts removed from the shoe and reinstall it on the cutter bar.

Mowing Cutter Bars **21**

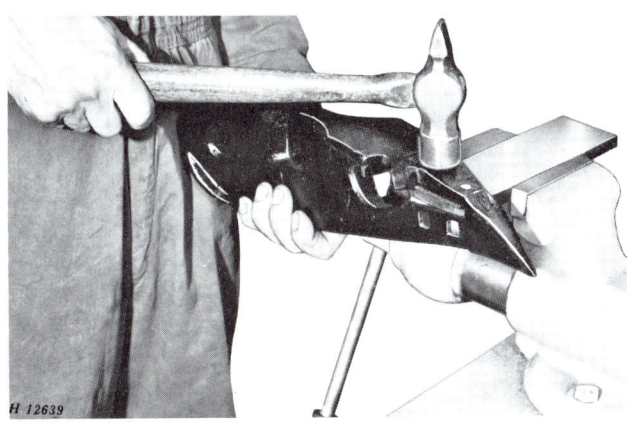

Fig. 59 — Replacing Ledger Plate on Outer Shoe

ROTARY-KNIFE CUTTER BARS

Rotary-knife cutter bars cut crops differently than the reciprocating knives discussed so far. They cut faster by spinning sharp, steel knives into the stand of crops (Fig. 60).

Oval disks are mounted on short, splined drive shafts sticking up from the mechanical drive in the cutter bar. Each disk has a knife bolted to each side (Fig. 61). When the cutter drive spins the disks, the knives whirl around cutting the crops.

Fig. 60 — Mower With A Rotary-Knife

Fig. 61 — Knives On The Oval Disks

Little maintenance, other than checking oil, greasing a few fittings, and changing dull knives, is required.

Adjust height and knife pitch angle of rotary cutter bars to meet field conditions. Height can be varied between 2 and 6 inches (5 and 15 cm) to cut close and level and to clear rocks. It's done by moving the skid shoes on the ends of bar up or down (Fig. 62).

Fig. 62 — Skid Shoe

Adjust the pitch angle by changing the length of the turnbuckles on each side of the machine (Fig. 63).

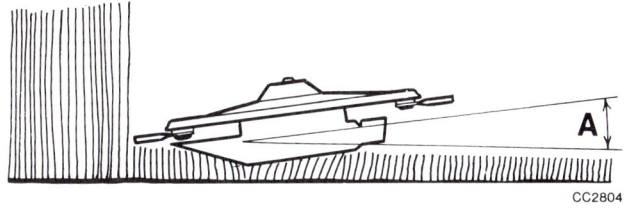

Fig. 63 — Pitch Angle

If the height and pitch angle are adjusted, you may want to adjust the safety curtains too.

Also, adjust the platform so weight is distributed properly (Fig. 64). Turn the load springs so about 25 pounds (11 kg) more weight is on the gear sheave.

FOS — 56 Litho in U.S.A.

22 Mowing Cutter Bars

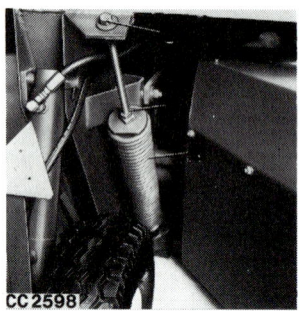

Fig. 64 — Adjust Weight Distribution

The oval disks must be set at right angles to one another so rocks pass through (Fig. 65).

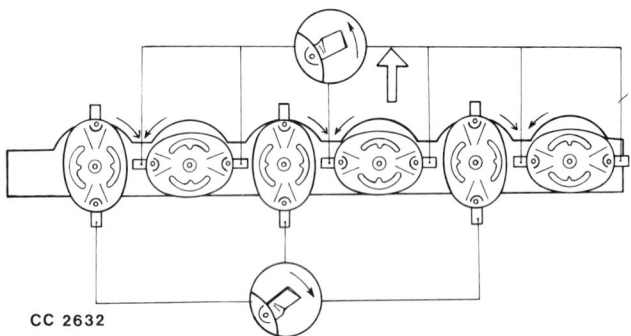

Fig. 65 — Disks At Right Angles

Read your operator's manual for more details of adjustments, maintenance, and operation.

SUMMARY

For good operation, the cutter bar must be properly adjusted and maintained, and parts must be replaced when necessary. Worn parts cannot do their job efficiently and continued operation can cause more damage. *It never pays to try to get by with worn parts.*

Proper adjustment is probably the most important factor in long cutter bar life with a minimum of repair. If a cutter bar is not kept in adjustment, repair and replacement will result in costly maintenance.

To summarize: Maintenance of the cutter bar consists of keeping the knife sharp, repairing or replacing worn or damaged parts, and lubricating as recommended by the manufacturer.

Repair or replacement of parts is often a matter of judgment. When in doubt whether to repair or replace, always replace the parts in question. This avoids unnecessary breakdowns.

TEST YOURSELF
QUESTIONS

1. What is the primary disadvantage of pitman-type drives?

2. What are the functions of the knife hold-down clips and wear plates?

3. What is cutter-bar lead?

4. Name two basic methods of establishing correct cutter-bar lead.

5. In what sequence should knife sections be replaced when replacing all sections?

6. Why is it necessary to shear knife section rivets off rather than punch them out?

7. What is the recommended procedure for removing a knife?

(Answers on back page.)

SPRAYING NOZZLES / PART 2

Fig. 1 — Spraying Nozzles At Work

INTRODUCTION

Spraying nozzles can apply an accurate amount of liquid uniformly over a large area.

To do this, a nozzle must atomize a liquid into small droplets. Then it must spray these droplets in a specific pattern at a particular flow rate.

The nozzle does four basic jobs:

1. Meters liquids at a certain flow rate.

2. Atomizes liquids into droplets.

3. Disperses the droplets in a specific pattern.

4. Propels the droplets for proper impact.

Let's examine these four jobs in terms of nozzle *flow* rate and *atomization.*

First — flow rate.

FLOW RATE

The nozzle flow rate is chiefly controlled by the metering passages. However, pressure, density, and viscosity also affect the flow rate.

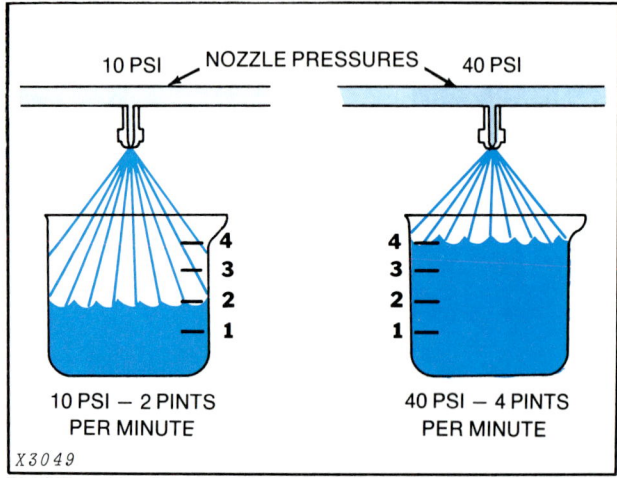

Fig. 2 — Flow Rate Is In Proportion To Pressure

With most nozzles, flow rate is in proportion to pressure. **The higher the pressure, the higher the flow rate.**

To double the flow rate, the pressure must be increased four times (Fig. 2). Conversely, to decrease the flow rate by half, the pressure must be reduced to one-fourth of the original pressure.

Density and viscosity of the fluid can affect the flow rate in different ways. Under some conditions the flow rate will actually increase with more viscosity. Usually, however, **the more dense and viscous the liquid, the lower the flow rate.**

ATOMIZATION

Atomization is the liquid break-up caused by the tearing action of air. As the liquid exits from the nozzle, it is in unstable sheets or jets, which collapse into tiny droplets. Usually the finer droplets are toward the inside of the spray pattern, while the larger droplets are at the outer edge.

The size of droplets is affected by nozzle rating, pressure, viscosity, and surface tension.

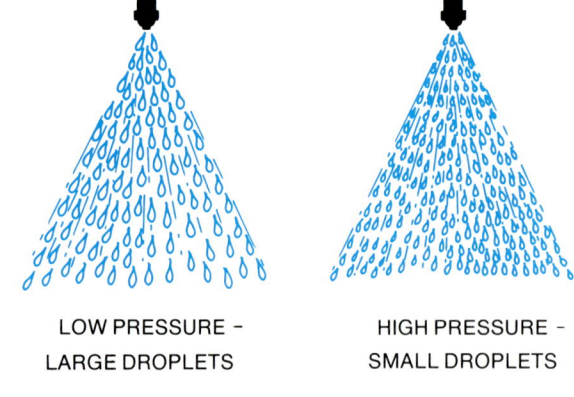

Fig. 3 — Droplet Size Decreases As Pressure Increases

As pressure is increased, the droplet size decreases for a finer spray (Fig. 3). A limit is eventually reached where more pressure has little effect on finer atomization.

With more viscous or dense fluids, larger droplets occur. Usually higher pressures are required to break up the liquid into the desired spray.

While surface tension does not affect atomization as much as viscosity, this tension can be difficult to break up if the liquid is denser. Another liquid of the same viscosity but with less surface tension can be atomized more easily.

In some spray applications, such as insecticides, a finer droplet size is required in order to obtain good spray impact and coverage of the plant. But with greater atomization, the spray is more susceptible to wind drift. **Therefore, protective clothing and a respirator must be worn by the operator.**

Larger droplets are generally recommended for spraying herbicides. Tips with larger orifices such as flooding tips and lower operating pressures 15 to 50 psi (100 to 345 kPa) will give larger droplets and reduce harmful spray drift.

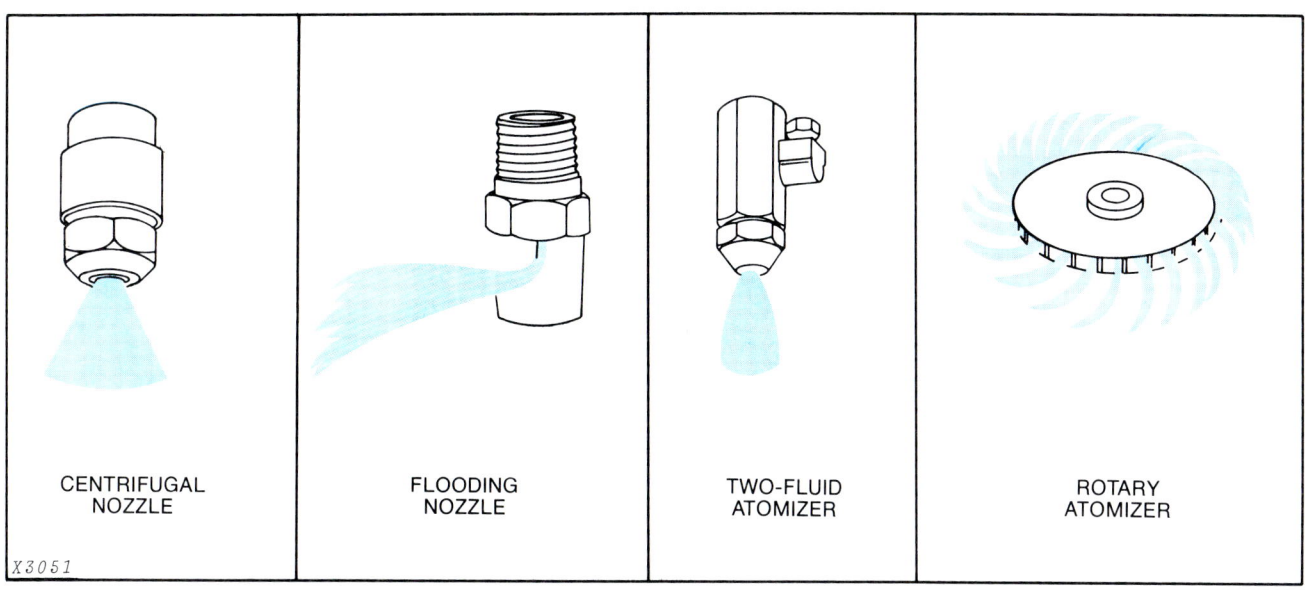

Fig. 4 — Common Types Of Nozzles

TYPES OF NOZZLES

Spraying nozzles are made in several designs. Each design atomizes fluids in a specific way to meet the job requirements.

The common types of nozzles are:

- **Centrifugal nozzles**
- **Flooding nozzles**
- **Two-fluid atomizers**
- **Rotary atomizers**

CENTRIFUGAL NOZZLES are most common (Fig. 4). They are made with a wide range of spray angles, spray patterns, and capacities.

FLOODING NOZZLES produce a spray in the shape of a fan or sheet (Fig. 4). They may also be called "fan-spray" nozzles.

TWO-FLUID ATOMIZERS can produce very fine droplets and will handle dense fluids. However, they require more pressure than other nozzles.

ROTARY ATOMIZERS are for big jobs; spraying thousands of gallons per hour by centrifugal force in a 360-degree spray pattern.

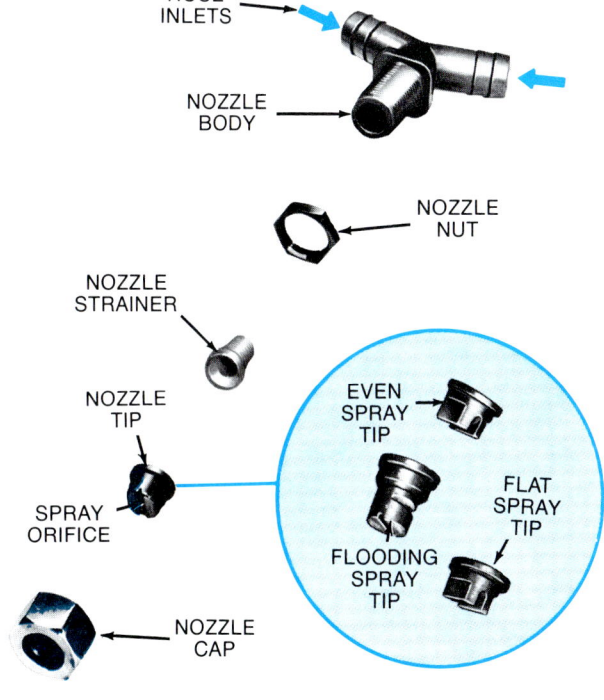

Fig. 5 — Typical Spraying Nozzle Assembly

Since we are concerned mainly with centrifugal pressure nozzles and flooding nozzles, we will discuss them in more detail later.

26 Spraying Nozzles

PARTS OF A NOZZLE

Nozzles have four basic components (Fig. 5).

- **Body**
- **Strainer**
- **Tip**
- **Cap**

Let's look at each part in detail.

NOZZLE BODY AND CAP

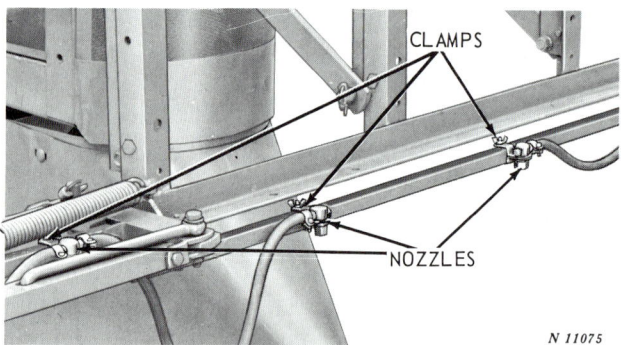

Fig. 6 — Nozzles Clamped To Dry Boom

Nozzle bodies and caps are usually made of brass, aluminum, stainless steel, zinc-plated steel, or nylon. These bodies and caps are designed to accept several different tips that spray various patterns (see Fig. 5). Usually the nozzle body is attached to a dry boom or a wet boom. On a dry boom, the nozzle bodies are clamped to the boom and the nozzles are supplied with fluid by hoses (Fig. 6).

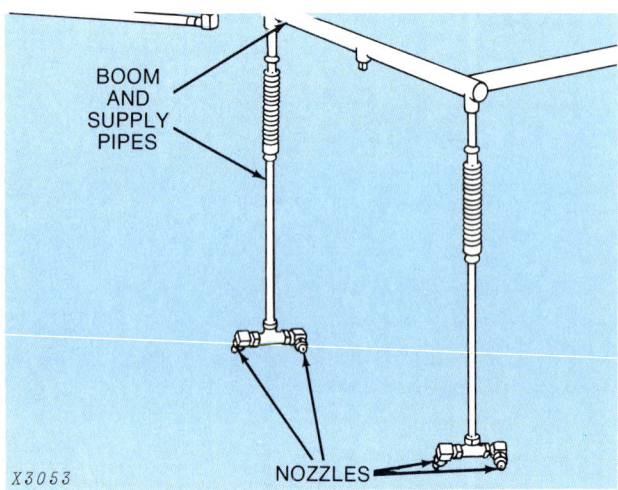

Fig. 7 — Nozzles Installed In A Wet Boom

On a wet boom, the nozzles are attached to a pipe which carries the fluid to each nozzle while functioning as part of the boom structure (Fig. 7).

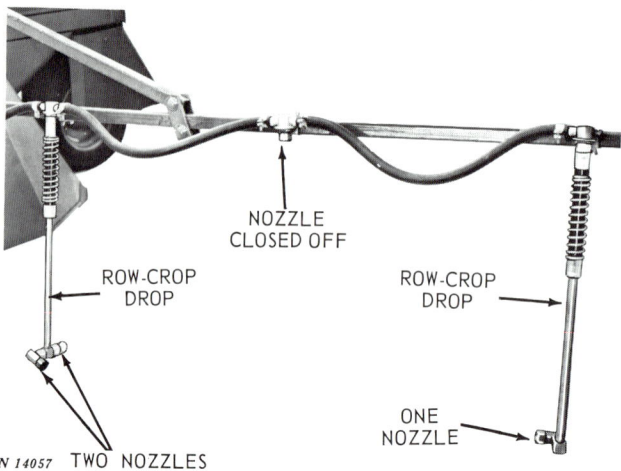

Fig. 8 — Swivel Nozzles On Row-Crop Drops

Another kind of nozzle body is the *swivel type*. Swivel bodies are attached to the lower end of a row-crop drop. This allows more accurate placing of spray for use after the crop has emerged, such as for insecticide spraying of the plants. A double swivel nozzle is used between the rows to direct the spray in both directions while a single swivel nozzle is used at the outer ends of the boom.

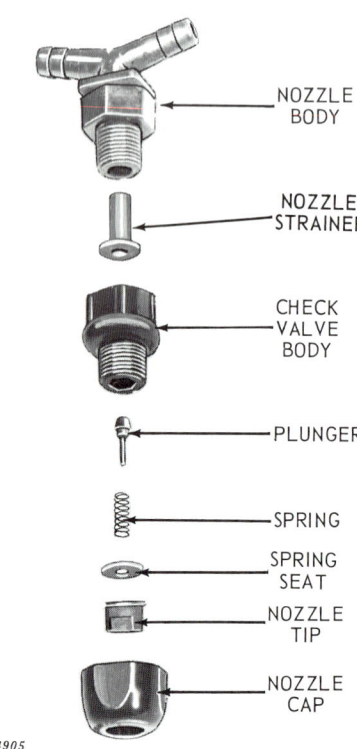

Fig. 9 — Plunger-Type Nozzle Check Valve

FOS — 56   Litho in U.S.A.

In some spraying operations, it is nice to have a quick shut-off at each nozzle so the nozzles won't drip at row ends or grass waterways. Check valves can be placed in the nozzle body (Fig. 9) to stop the flow.

When the line pressure drops below a certain low pressure, the valve automatically shuts off all flow. There are three valves used to do this: the plunger-type (Fig. 9), the ball-type, and the diaphragm-type check valve.

The plunger and diaphragm check valves are contained in their own body and attach to the nozzle body. However, the ball check valve can be used in place of the tip strainer.

NOZZLE STRAINERS

Nozzle screens or strainers (Fig. 10) provide the last screening of foreign material in the spraying system. This helps to prevent clogging at the nozzle tip.

On most *screens,* the screen material is stainless steel with mesh sizes of 50, 100, or 200 openings per linear inch (0.25 mm). The screen assembly is normally cylindrical and fits inside most nozzle bodies. This type of nozzle screen is used more with liquid-concentrate chemicals. When using powder chemicals, which are abrasive, a slotted *strainer* should be used (Fig. 10, right). These strainers have a larger mesh size equivalent to 16, 25, or 50 mesh.

When selecting a nozzle screen, choose the screen which has a mesh opening just a little *smaller* than the orifice in the nozzle tip. The smaller or finer the mesh, the more likely it is to clog.

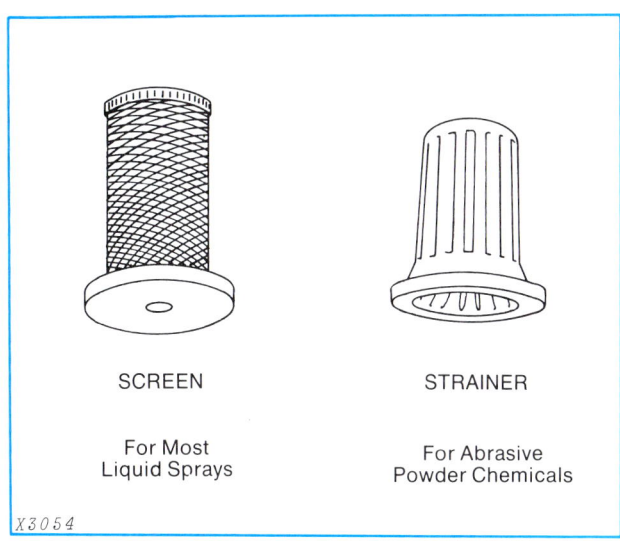

Fig. 10 — Nozzle Strainers

A *clean* screen or strainer is vital for efficient and accurate distribution of the spray material. They should be checked and cleaned often. If not, erratic spray patterns, improper metering and delivery, or complete clogging of the nozzle will occur.

NOZZLE TIPS AND SPRAY PATTERNS

Centrifugal and flooding nozzles have tips which provide five basic spray patterns (Fig. 11).

The centrifugal pressure nozzle produces a hollow or solid cone spray pattern. The flooding nozzle produces flat, even, or flooding spray patterns. With all these nozzles, various flow rates and spray angles are available.

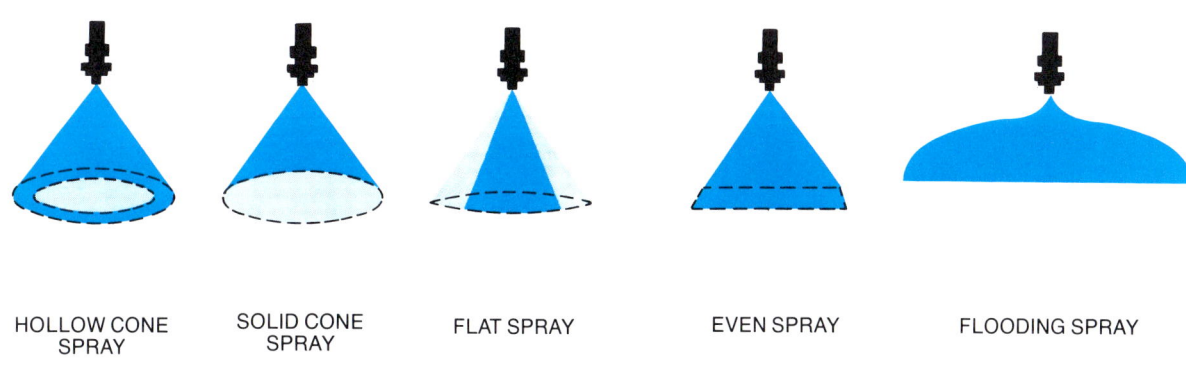

Fig. 11 — Types Of Spray Patterns

28 Spraying Nozzles

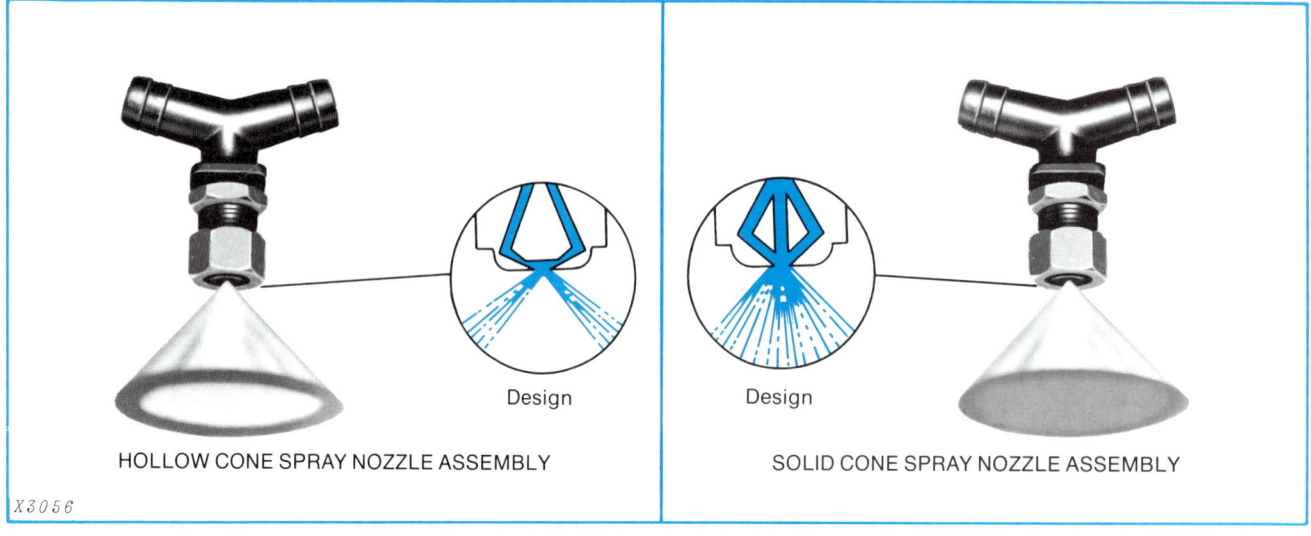

Fig. 12 — Hollow And Solid Cone Spray Pattern Nozzles

Hollow And Solid Cone Spray

Hollow and solid cone nozzles (Fig. 12) are popular for spraying high-volume fungicides, insecticides, and herbicides on row-crops (Fig. 13). These nozzles are generally used on spray booms or spray guns. They are more resistant to clogging from abrasive, wettable powders than other nozzles.

Fig. 13 — Row-Crop Spraying With Cone-Type Nozzles

These nozzles spray a cone pattern with large droplets at low pressures. The spray angle may be from 30 to 120 degrees. Hollow cone nozzles (Fig. 12) produce a more uniform, finely atomized spray than solid cone nozzles.

Notice in Fig. 12 how the spray pattern is achieved. Hollow cone nozzles contain a core that swirls the liquid around before it exits the orifice. The solid cone nozzle has a center passage that supplies a jet of liquid which fills in the hollow cone.

Flat And Even Spray

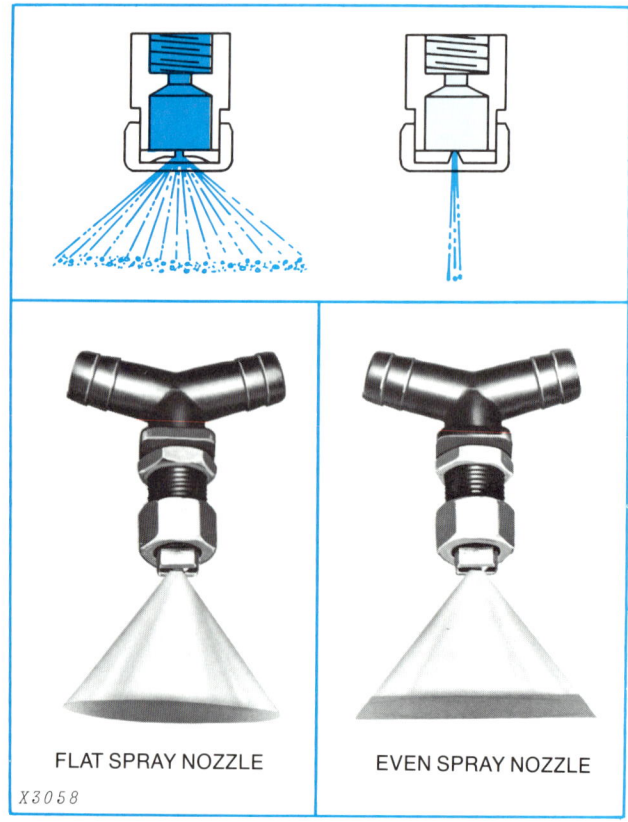

Fig. 14 — Fan Spray Pattern Nozzle

Flat and even spray patterns are produced by a fan-spray nozzle (Fig. 14).

The **flat** spray pattern is used for applying fertilizers, herbicides, and insecticides. It may also be used on boom sprayers, broadcast sprayers, planters, and cultivators.

FOS — 56   Litho in U.S.A.

Spraying Nozzles **29**

Fig. 15 — Overlapping Boom Spraying

This nozzle sprays a fan-shaped pattern with a gradually tapered edge. The spray angle may be between 65 and 80 degrees. When this pattern is properly overlapped, it provides even distribution of the liquid (Fig. 15).

Fig. 16 — Band Spraying

The **even** spray pattern is used for applying liquid in an even band (Fig. 16). Often it is used with planter attachments for row patterns.

The even spray gives a fan-shaped pattern with uniform distribution across its full spray width. The spray angle may be between 80 and 95 degrees.

Flooding Spray

Flooding nozzles deliver a wide, flat spray with large droplets (Fig. 17). As the liquid discharges from a plain orifice, it strikes a curved deflector which deflects the spray about 75 degrees. The spray angle may be between 70 and 160 degrees. These nozzles can be mounted in various positions to provide different patterns.

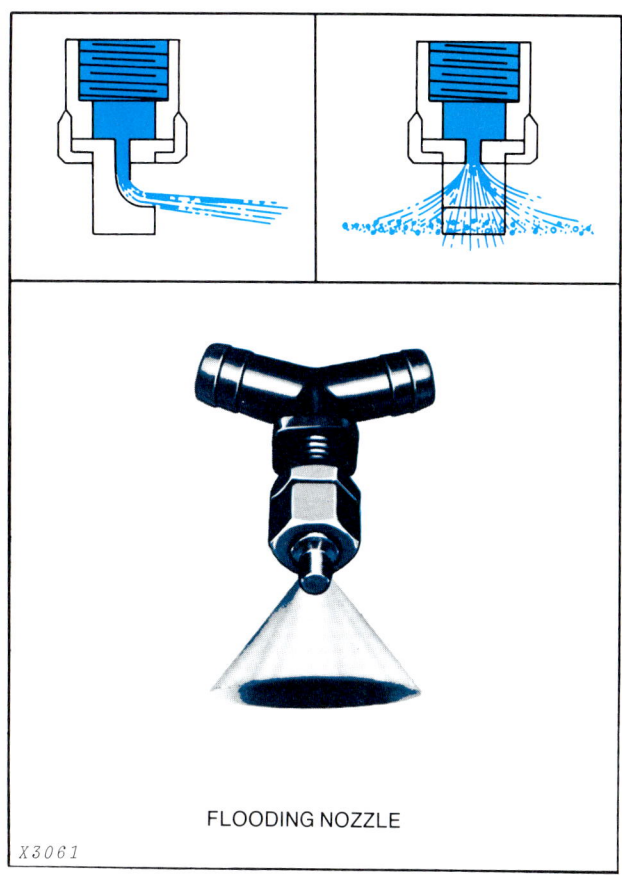

Fig. 17 — Flooding Spray Pattern Nozzle

Fig. 18 — Boomless Broadcast Spraying

Flooding nozzles are often used for broadcast application of fertilizers and defoliants. Also, they are used in post-emergence application of herbicides. Usually these nozzles are mounted on boom sprayers or lay-by rigs. Sometimes they are used alone as in boomless broadcast spraying (Fig. 18).

FOS — 56   Litho in U.S.A.

OTHER TYPES OF SPRAY EQUIPMENT

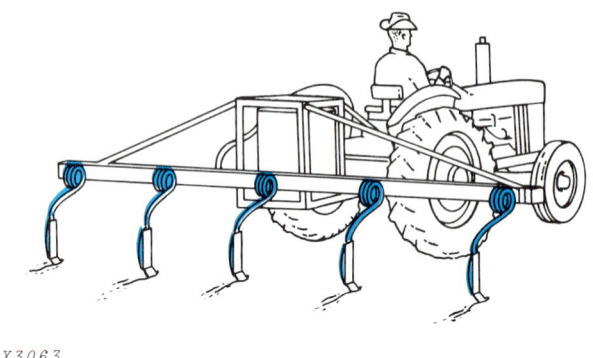

Fig. 19 — Subsoil Application Of Liquid Fertilizer

Special equipment is available for other application methods, such as applying liquid fertilizer with subsoil attachments (Fig. 19).

These attachments carry the liquid into the soil under pressure. Nozzle check valves are normally used to prevent spray from dripping when turning at row ends.

Dribble applicators and wide-spray jets are also available.

Fig. 20 — Hand Spraying Gun

Hand spray guns are used for selective spraying (Fig. 20).

These guns can be used for fence rows, small weed patches, and for insect control in barns.

The hand gun may be used alone or connected to a mounted spray boom. Most guns contain their own check valves.

SELECTION OF NOZZLES

Select nozzles that produce the proper droplet size and have an application rate within the recommended range of pressure.

Many organizations provide handy charts and calculators to help people pick nozzle tips. There is a picture of one on page 35! The calculator puts the factors of spraying into a handy, usable tool.

Nozzle tips are made of: brass, aluminum, nylon, stainless steel, hardened stainless steel, chrome plated brass, and brass with a tungsten carbide orifice.

Brass and aluminum tips are the cheapest, but the metal is soft and the tips wear fast. Tips made from harder metals cost more but they last longer.

As nozzle tips wear, the rate of application increases. Tests have shown that some wettable powders may wear nozzle tips sufficiently to increase the rate as much as 12 percent after spraying 50 acres. For this reason, frequent calibration of equipment is advisable.

CALIBRATING THE APPLICATION RATE

To perform accurately in the field, the flow rate of sprayer nozzles must be known. Let's look at suggestions for calibration.

As we have mentioned, flow rate is affected by nozzle passages, tip design, pressure, viscosity, and density.

Since most application rate charts are based on the flow characteristics of water, the sprayer must be calibrated for the addition of chemicals. This calibration also checks for variations in nozzle output and in equipment performance.

Many studies have determined that volume and pattern may vary considerably, particularly when wettable powders are used. Therefore, the sprayer should be calibrated daily, and the operating pressure should be gradually lowered to compensate for wear until the nozzle tips require replacement.

Fig. 21 — Collecting Spray From Each Nozzle

There are several calibrating methods; however, the method which follows is simple and one of the most direct means of calibrating. The only measuring devices necessary are:

1) Stop watch

2) Container graduated in fluid ounces (milliliters) or quarts (Liters)

3) Tape measure

Precalibrating

Calibration is required because of worn nozzle tips or line pressure loss. Check the gallons per acre (L/ha) flow ONLY AFTER setting proper pressure and installing the proper nozzle tips as indicated by a nozzle tip calculator or your operator's manual.

Check the nozzle tips (unless they are all new and of the correct size) as follows:

1. Install clean water in the tank.

2. With sprayer stationary, operate the sprayer at pressure setting indicated by nozzle tip calculator. Check for leaks and spray pattern from the nozzles.

3. Collect the spray from each nozzle tip for the same period of time (15-30 seconds). Record sample from each tip and compare with average.

4. Replace tips having an uneven spray pattern and/or where tip flow is greater or less than 10% of average.

SAFETY RULES

CAUTION: Agricultural chemicals can be dangerous. Improper use can injure persons, animals, plants, soil, or other property. Handle and apply with care.

When mixing, calibrating, or working around chemicals, use the following equipment and supplies:

• Protective clothing; cap, gloves, respirator, goggles, footwear, etc.
• Clean water supply
• Detergent

If spray gets on your body, wash IMMEDIATELY with clean water and detergent.

Select an area where you can safely fill, flush, calibrate, and decontaminate the sprayer without pesticides drifting or running off to contaminate people, animals, vegetation, or a water supply. Select an area where it will be impossible for children to come into contact with pesticides.

To reduce spray drift hazards:

• **Use large nozzle tips operated at lower pressures.**
• **Do not operate sprayer at pressures over 50 psi (345 kPa).**
• **Do not make spray applications when winds exceed 10 mph (16 km/h).**

32 Spraying Nozzles

HOW TO CALIBRATE

After selecting the proper nozzle you will want to calibrate the spray rig to make sure it sprays the right volume and rate.

1. Fill the sprayer with properly mixed chemical solution. In this example, let's assume we have a chemical with a specified application rate of 8 gallons per acre (75 liters per hectare).

2. Measure calibration distance in the field; see chart for the required distance for a given row spacing for broadcast spraying or for a given band width for band spraying. Drive the sprayer across the marked distance at operating speed and with initial sprayer pressure. Write down the **travel time, throttle setting,** and **spray pressure.** Determine the operating speed in miles per hour (mph) or kilometers per hour (km/h). Use either of the following formulas to calculate speed.

$$mph = \frac{0.682 \times \text{length of run (ft.)}}{\text{time in seconds}}$$

$$km/h = \frac{3.6 \times \text{length of run (m)}}{\text{time in seconds}}$$

3. For this example, assume you have calculated a speed of 8 mph (13 km/h) with an enitial pressure of 32 psi (220 kPa) in step 2.

Now, with the sprayer parked and the parking brake set, turn on the spraying system. Run it at the same throttle setting and spray pressure you used in step 2.

Collect 2 pints (1 L) of spray from one nozzle and write down the number of seconds required to collect it. In this example; **60 seconds.**

Use the following formula to determine gallons (liters) per minute per nozzle.

$$gpm/nozzle = \frac{8}{\text{seconds/pint}}$$

$$gpm/nozzle = \frac{8}{30}$$

$$gpm/nozzle = 0.27 \text{ gpm}$$

or in metrics:

$$L/min/nozzle = \frac{60}{\text{seconds/L}}$$

$$L/min/nozzle = \frac{60}{60}$$

$$L/min/nozzles = 1 \text{ L}$$

FOS — 56   Litho in U.S.A.

4. Now use the following formula to determine what the gallons per acre (liters per hectare) **application rate** will be at the predetermined speed and sprayer pressure.

$$gpa = \frac{5{,}940 \times gpm/nozzle}{\text{speed (mph)} \times \text{nozzle spacing}}$$

$$gpa = \frac{5{,}940 \times 0.27}{8 \text{ mph} \times 20 \text{ inches}}$$

$$gpa = \frac{1603.8}{160}$$

$$gpa = 10$$

or in metrics:

$$L/ha = \frac{600{,}000 \times L/min/nozzle}{\text{speed (Km/h)} \times \text{width (nozzle spacing in mm)}}$$

$$L/ha = \frac{600{,}000 \times 1}{13 \times 500 \text{ mm}}$$

$$L/ha = \frac{600{,}000}{6{,}500}$$

$$L/ha = 92.3$$

5. If the calculated gpa (L/ha) does not agree with the desired application rate, adjust the pressure setting and collect spray again. Remember: raising the pressure increases the application rate; lowering the pressure decreases the rate. To double the rate, the pressure must be increased four times. Use nozzles with higher flow rates rather than increase pressure of the sprayer beyond the manufacturer's recommendations.

6. In this example, 8 gpa (75 L/ha) is the desired application rate. So, reduce the pressure down to 22 psi (150 kPa). Then it will take 75 seconds to collect 2 pints (1 liter) of chemical from one nozzle. Again, use the formula:

$$gpm/nozzle = \frac{8}{75 \text{ seconds/pint}}$$

$$gpm/nozzle = \frac{8}{37.5}$$

$$gpm/nozzle = 0.21$$

Or in metric:

$$L/min./nozzle = \frac{60}{75 \text{ seconds/liter}}$$

$$L/min./nozzle = 0.8$$

7. Use the formula to determine application rate in gallons per acre (gpa) or liters per hectare (L/ha).

$$\text{gpa} = \frac{5940 \times .21 \text{ gpm/nozzle}}{8 \times 20}$$

$$\text{gpa} = \frac{1247.4}{160}$$

$$\text{gpa} = 7.8$$

or in metric:

$$\text{L/ha} = \frac{600{,}000 \times 0.8}{13 \times 500}$$

$$\text{L/ha} = \frac{480{,}000}{6{,}500}$$

$$\text{L/ha} = 74$$

The sprayer is now spraying at the required rate of 8 gpa (75 L/ha).* Several manufacturers provide handy calculators to do the arithmetic.

CALIBRATION DISTANCE CHART

Rows		Bands	
Row Spacing (inches [m])	Distance (feet [mm])	Band Width (inches [m])	Distance (feet [mm])
40 (1.02 m)	102 (3,111 mm)	14 (0.36 m)	292 (8,963 mm)
38 (0.97 m)	107 (3,263 mm)	12 (0.31 m)	340 (10,426 mm)
36 (0.91 m)	113 (3,446 mm)	10 (0.25 m)	408 (12,443 mm)
34 (0.86 m)	120 (3,659 mm)		
32 (0.81 m)	128 (3,903 mm)		
30 (0.76 m)	136 (4,146 mm)		
28 (0.71 m)	146 (4,450 mm)		
26 (0.66 m)	157 (4,785 mm)		
24 (0.61 m)	170 (5,181 mm)		
22 (0.56 m)	186 (5,668 mm)		
20 (0.51 m)	204 (6,216 mm)		

NOZZLE ADJUSTMENT AND CARE

Proper nozzle adjustment and care is required to maintain good nozzle performance.

The distance from the nozzle tips to the sprayed surface, the angle of spray, and the spacing of the nozzles all have a bearing on proper spray coverage. Nozzles must be spaced equally along the spray boom and positioned correctly in relation to the row. On wet booms, the nozzle position is fixed and cannot be changed. However, on dry booms the nozzles are clamped to the boom and can be moved.

PROBLEMS WITH NOZZLES

The illustrations which follow show the most common problems with booms and flat spray nozzles. Similar problems may occur with other types of nozzles.

Worn Or Plugged Nozzles

Fig. 22 — Worn Or Plugged Nozzles

Worn or plugged nozzles will result in untreated strips as seen in Fig. 22.

Clean nozzles periodically; particularly when applying wettable powders. A worn nozzle tip may result in over-application and many times cannot be detected by observation. Repeated calibration of the sprayer will detect worn nozzle tips.

If a nozzle cloggs or some other part malfunctions, shut down the sprayer and pump, and release pressure from system.

Never touch nozzle tips or other sprayer parts to your lips to blow out trash. Have spare tips available for replacement.

Misaligned Nozzles

Fig. 23 — Misaligned Nozzles

If flat type nozzle tips are not aligned parallel with the boom, coverage will not be complete (Fig. 23). Wrench flats are provided so that the tips can be properly aligned.

Improper Nozzle Tips

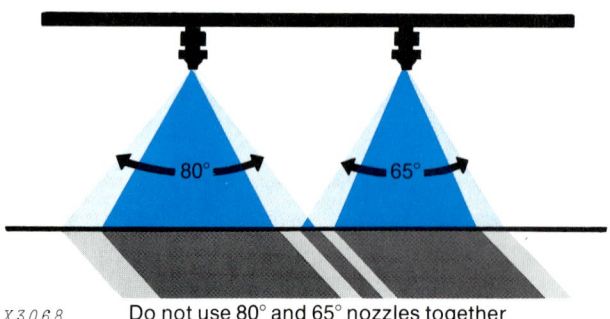

X3068   Do not use 80° and 65° nozzles together

Fig. 24 — Improper Use Of Nozzle Tips

Nozzle tips of different sizes and angles must NOT be used on the same boom (Fig. 24). Irregular spray coverage will result.

Spray Boom Not Level

X3069

Fig. 25 — Spray Boom Not Level

If the boom is not level, coverage will be irregular (Fig. 25).

Spray Boom Too Low

X3070

Fig. 26 — Spray Boom Too Low

If the boom is carried too high or too low, uneven patterns and coverage will occur (Fig. 26). Follow the manufacturer's recommendations for height.

CLEANING NOZZLES

Clean nozzles. Failure to keep them clean may result in irregular spray patterns and plugged nozzles.

Disassemble the nozzle and clean the parts with a soft brush, toothbrush, or toothpick and clean water or safe solvent. **Do not use metal probes of any kind to clean tips and screens.**

At the end of the season, remove and clean nozzle tips and screens and store them in a jar of diesel fuel.

TEST YOURSELF

QUESTIONS

1. Name three of the four basic functions of a nozzle?

2. Fill in blanks with "faster" "slower": The higher the nozzle pressure, the .... the flow rate.

3. True or false? "The higher the nozzle pressure, the finer the spray."

4. How many times must the pressure be multiplied to **double** the flow rate?

5. What is used in a nozzle for quick shut-off?

6. True or false? "As nozzles wear out, the flow rate normally decreases."

7. Which of the following may be used to clean nozzles.

    a) Toothpick   c) Metal probe

    b) Wire brush   d) Toothbrush

8. Why should a sprayer be calibrated?

ANSWERS

1. The four basic functions of a nozzle are: metering liquid, atomizing liquid into droplets, dispersing the droplets in a specific pattern, and propelling the droplets for proper impact. (Any three for correct answer.)

2. "Faster"

3. True.

4. *Four* times.

5. A *check valve.*

6. False.

7. Only the toothpick (a) and the toothbrush (d) are safe for cleaning spray nozzles. Metal probes or wire brushes can damage the nozzles and their orifices.

8. It is necessary to calibrate a sprayer because nozzle flow rate charts are based on the flow characteristics of water alone. Calibration also checks variations in nozzle output and in equipment performance.

ANSWERS FOR QUESTIONS ON PAGE 22

1. Pitman-type drives usually vibrate more than pitmanless drives, and cannot be operated as fast due to the hammering effect of the pitman thrust.

2. The knife hold-down clips and wear plates *hold the knife in proper position* for the best cut.

3. *Cutter bar lead* is an adjustment that aligns the cutter bar with the pitman when the pressure of the crop pushes the cutter bar back.

4. The two methods for establishing cutter bar lead are the parallel-line method and the straight-board method.

5. When replacing all knife sections, *alternate* sections should be removed and replaced. Then the remaining sections can be replaced in the same manner. This eliminates rivet-shift and keeps the knife straight.

6. If the rivets are punched out, the rivet holes in the knife back may be enlarged. This will permit the sections to shift and cause rivet failure.

7. First block up the cutter bar. Remove the drive arm or pitman. Grasp the knife by its back side with *both hands* and pull the knife out. Maneuver any sections that hang up with the blocks. This avoids placing your fingers in a vulnerable position.

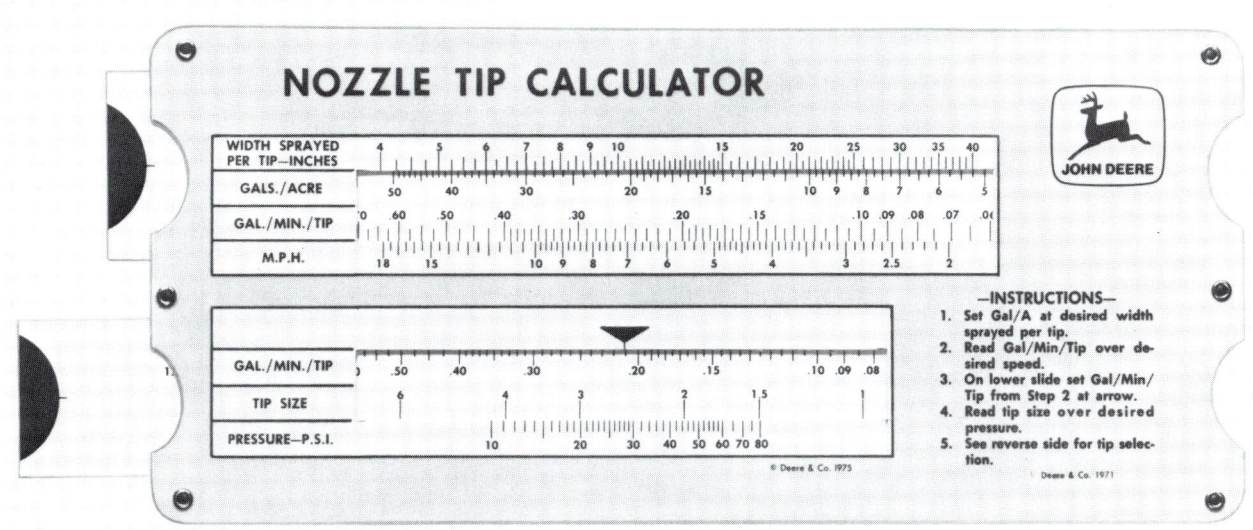